Modulation of Protein Stability in Cancer Therapy

Kathleen Sakamoto • Eric Rubin

Editors

Modulation of Protein Stability in Cancer Therapy

 Springer

Editors
Kathleen Sakamoto
David Geffen School of Medicine
Los Angeles, CA
USA
ksakamoto@mednet.ucla.edu

Eric Rubin
Cancer Institute of New Jersey
New Brunswick, NJ
USA
eric-rubin@merck.com

ISBN: 978-1-4419-2401-8   e-ISBN: 978-0-387-69147-3
DOI: 10.1007/978-0-387-69147-3
Springer Dordrecht Heidelberg London New York

Printed on acid-free paper

Springer is part of Springer Science+Business Media (www.springer.com)

Preface

Over the past two decades, there has been a tremendous increase in our understanding of ubiquitination and proteasome degradation. As the editors, we thank Springer Publishing for allowing us to organize this collection of reviews on the ubiquitin-proteasome system, oncogenesis, and cancer therapy. We asked our colleagues, who are experts in the field, to provide an overview of their research and recent progress. Each chapter covers a broad range of topics that include defects in ubiquitination identified in specific tumors to new directions to treat cancer. In the Introduction, Rati Verma gives the background of our current understanding of the proteasome. A general overview of ubiquitin ligases and cancer is provided by Angelika Burger and Arun Seth. Patricia McChesney and Gary Kupfer discuss the role of the Fanconi anemia/BRCA1 pathways in breast cancer and potential targets for therapy. Defects in the tumor suppressor, von Hippel–Lindau E3 ligase, and the role of this protein complex are discussed by William Kim and William Kaelin. Kyung-bo Kim and colleagues describe the development of novel proteasome inhibitors to treat cancer patients. Progress in our understanding of deubiquitinating enzymes is summarized by Massimo Loda and his colleagues. Finally, Agustin Rodriguez-Gonzalez and Kathleen Sakamoto discuss an approach to recruit cancer-causing proteins to ubiquitin ligases through a chimeric molecule known as protacs. We hope that this collection of reviews will provide examples of how discoveries in basic sciences have provided new insights into our understanding of the pathogenesis of specific tumors and possible new therapeutic approaches to treat cancer.

Los Angeles, CA   Kathleen Sakamoto
New Brunswick, NJ   Eric Rubin

Contents

Contributors

Christopher L. Brooks
Institute for Cancer Genetics, and Department of Pathology, College of Physicians & Surgeons, Columbia University, New York, NY, USA

Angelika M. Burger
Department of Pharmacology & Experimental Therapeutics, Greenebaum Cancer Center, University of Maryland School of Medicine, Baltimore, MD, USA

Wei Gu
Institute for Cancer Genetics, and Department of Pathology, College of Physicians & Surgeons, Columbia University, New York, NY, USA

Yik Khuan Ho
Department of Pharmaceutical Sciences, College of Pharmacy, University of Kentucky, Lexington, KY, USA

William G. Kaelin, Jr., M.D.
Dana-Farber Cancer Institute, Boston, MA, USA

Kyung-Bo Kim
Department of Pharmaceutical Sciences, College of Pharmacy, University of Kentucky, Lexington, KY, USA

William Y. Kim, M.D.
University of North Carolina School of Medicine, Lineberger Cancer Center, Chapel Hill, NC, USA

Gary Kupfer
Department of Pathology, Yale University School of Medicine, New Haven, CT, USA

Massimo Loda
Departments of Medical Oncology and Pathology, Dana Farber Cancer Institute and Brigham and Women's Hospital, Harvard Medical School, Boston, MA, USA

Patricia McChesney
Section of Pediatric Hematology-Oncology, Department of Pediatrics,
Yale University School of Medicine, New Haven, CT, USA

Pooja Pungaliya, Ph.D.
Postdoctoral Fellow, Biological Technologies, Wyeth Research, Cambridge,
MA, USA

Agustin Rodriguez-Gonzalez
Department of Pediatrics, Department of Pathology & Laboratory Medicine,
Gwynne Hazen Cherry Laboratories, Jonsson Comprehensive Cancer Center,
Los Angeles, CA, USA

Eric Rubin, M.D.
Vice President, Oncology Clinical Research, Merck Research Laboratories, USA

Arun K. Seth
Laboratory of Molecular Pathology, Department of Anatomic Pathology,
Sunnybrook Research Institute, University of Toronto, Toronto, ON, Canada

Rati Verma
Department of Biology, Howard Hughes Medical Institute,
California Institute of Technology, Pasadena, CA, USA

Marie Wehenkel
Department of Pharmaceutical Sciences, College of Pharmacy,
University of Kentucky, Lexington, KY, USA

Ubiquitin Ligases and Cancer

Angelika M. Burger and Arun K. Seth

Abstract Ubiquitin E3 ligases are the most prevalent cancer genes besides protein kinases, suggesting a critical role in cancer development and progression. Two major classes of E3 ligases can be distinguished: the HECT-type and the adaptor-type E3s. The latter include RING-finger, PHD, and U-box domain ligases. The vast majority of the known and cancer-associated E3s belong to the RING-finger gene family. This chapter describes key representatives of each class of E3s, their substrates, and relevance to cancer.

Keywords Ubiquitin ligase • Ubiquitin • RING-type E3 ligase • HECT-type E3 ligase • Cancer

Introduction

Protein homeostasis at the cellular level is delicately balanced by de novo synthesis, posttranslational modification, and degradation processes. Degradation fulfills several functions in a cell: the elimination of damaged and unneeded proteins, activation of protein precursors by partial hydrolysis, and the complete proteolysis of a specific protein or the proteolysis of proteins that regulate functions and levels of others (Glickman and Ciechanover 2002; Ciechanover 2005). Two major intracellular proteolytic systems exist: a lysosomal and a non-lysosomal system (Bohley 1995). Lysosomal proteases that are localized in lysosomes and endosomes are responsible for the degradation of 10–20% of proteins in the mammalian cells. Lysosomal substrates are extracellular proteins, cell surface receptors, and intracellular organelles.

A.M. Burger (✉) and A.K. Seth
Department of Pharmacology and Experimental Therapeutics, and Greenebaum
Cancer Center, University of Maryland School of Medicine, 655 West Baltimore Street,
Baltimore, MD 21201, USA
e-mail: angelikaburger@aol.com

K. Sakamoto and E. Rubin (eds.), *Modulation of Protein Stability in Cancer Therapy*,
DOI: 10.1007/978-0-387-69147-3_1, © Springer Science+Business Media, LLC 2009

About 80% of cellular proteins are destructed in the ubiquitin-proteasome system (U-PS) (Bohley 1995; Goldberg et al. 1995; Lecker et al. 2006).

The knowledge of the molecular mechanisms governing the U-PS has exploded over the past decade and it has become obvious that ubiquitin tagging not only targets proteins for destruction, but that dependent on the lysine residue through which ubiquitin is linked to a protein, ubiquitination can have regulatory functions, including regulation of translation, DNA repair, activation of transcription factors and kinases, as well as endocytosis and the vesicular transport of membrane proteins (Weissman 2001; Hicke 2001).

The ubiquitin system is hierarchically structured involving three distinct classes of enzymes: E1 ubiquitin activating enzyme, E2 ubiquitin conjugating enzymes, and E3 ubiquitin ligases (Hershko et al. 1980; Hershko et al. 2000). Only two isoforms of the E1 ubiquitin activating enzyme exist indicating its essential function. E2 ubiquitin conjugating enzymes are manifold (~45) and are characterized by a 14–16 kDa core that is to be 35% conserved. In addition, they can bind to, for example, the human E6-AP COOH terminus (HECT) as well as the really interesting new gene (RING) domains of E3 ligases such as UBCH7. E2 are therefore believed to have a functional redundancy (Handley-Gaerhart et al. 1994; Weissman 2001; Burger and Seth 2004; Beckmann et al. 2004; Lecker et al. 2006).

E3 ubiquitin ligases are currently grouped into two major categories: the HECT-type and adaptor-type E3s, the latter containing a RING-finger, an U-box, or a plant homeodomain (PHD). The human genome contains over 400 ubiquitin ligase domain genes, which in their multitude confer specificity for protein substrates (Joazeiro and Weissmann 2000; Pray et al. 2002; Coscoy and Ganem 2003; Hatakeyama and Nakayama 2003; Beckmann et al. 2004; Burger and Seth 2004; Yang et al. 2004) (Fig. 1).

Many ubiquitin E3 ligases play a role in human diseases such as Alzheimer's disease, Parkinson's disease, and cancer. In this chapter, we will describe known ubiquitin E3 ligases in cancer (Table 1). Understanding their structure, similar catalytic mechanisms and various functions will provide us with unique targeting opportunities and biomarkers.

Ubiquitin E3 Ligases in Cancer

When the human genome was released, a census was conducted showing that protein kinase genes account for almost 10% of all currently known cancer genes (Futreal et al. 2004). A reexamination of the literature and available databases showed that E3 ubiquitin ligases are also the key mediators of tumorigenesis. Altogether, kinase and E3 genes represent more than 15% of the known cancer genes, underlining the importance of phosphorylation and ubiquitinylation signaling pathways in cancer development (Beckmann et al. 2004, 2005). In particular, 409 genes that encode putative domain-based E3 ubiquitin ligases were found in the human genome. Of these, respectively, 15 were among the 277 cancer genes identified by Futreal et al. (2004). Owing to the continuously emerging literature reporting correlations between

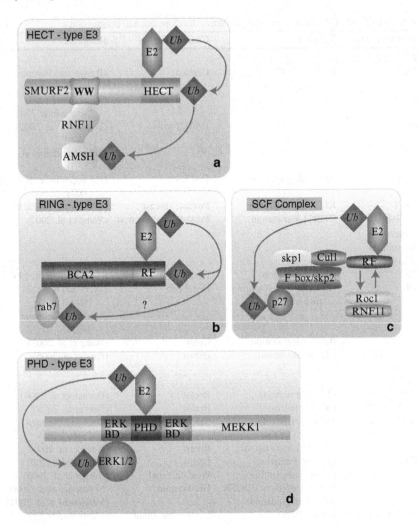

Fig. 1 Representatives of the major types of E3 ligases are shown. (**a**) The HECT-type E3 ligase Smurf2 has a WW domain and a HECT domain. The WW domain binds a co-activator protein such as RNF11 that mediates the recognition of the ubiquitinylation substrate, e.g., AMSH. The HECT domain binds the E2 conjugating enzyme and facilitates the ubiquitin (Ub) transfer onto the substrate via formation of a thioester intermediate with Ub. The RING-type E3 ligases can either be monomeric (**b**) such as human BCA2, or a multisubunit complex such as SCF (**c**). BCA2 binding to Rab7 and SCF acting on one of its prominent substrate, p27, are presented as examples. A hallmark of monomeric RING E3 ligases is autoubiquitination, hence Ub is transferred not only to the substrate by the E2/E3 complex, but also to the E3 RING-finger itself. The components of the SCF complex are labeled. Roc1 and RNF11 may have interchangeable function as RING-finger (RF) in the SCF complex. (**d**) MEKK1 represents the best-known PHD domain containing protein with E3 ligase activity. MEKK1 ubiquitinates ERK1/2 after binding to one of its ERK binding domains (ERK BD)

alterations in ubiquitinylation processes and tumorigenesis, the percentage of E3s involved in cancer development and progression is likely to further increase, and the E3 genes could serve as baits for the identification of additional cancer genes

Table 1 Ubiquitin E3 ligases with a role in cancer development and progression

Ubiquitin Ligase	Substrates	Cancer Type	Reference
HECT-type			
E6AP	Degrades p53 in presence of E6 oncoprotein	Cervical	(Brown and Pagano 1997)
			(Pagano and Benmaamar 2003)
Smurf2	AMSH, smad2 (degradation)	H&N, breast	(Fukuchi et al. 2002)
			(Subramaniam et al. 2003)
Smurf1	RhoA (degradation)	Pancreatic	(Loukopoulos et al. 2007)
WWP1	KLF5 (degradation)	Prostate, breast	(Chen et al. 2007a, b)
Nedd4-1	PTEN (degradation)	Prostate, endometrial	(Wang et al. 2007)
Adapter-type Multisubunit RING E3s			
SCF	p27 (degradation)	Breast	(Catzavelos et al. 1997)
VHL	Hif1-alpha (degradation under Normoxia)	Renal cell	(Tyers and Rottapel 1999)
			(Mani and Gelmann 2005)
APC/C	skp2^{scf} (degradation)	Colon	(Wang et al. 2003)
	p27 (stabilization)	Breast	(Park et al. 2005)
Monomeric RING E3s			
COP1	p53 (degradation)	Breast	(Dornan et al. 2004)
BRCA1/BARD1	p53 (degradation)	Breast	(Feki et al. 2005)
Mdm2	p53 (degradation)	Sarcoma, glioma	(Onel and Cordon-Cardo 2004)
Efp	14-3-3σ (degradation)	Breast	(Horie et al. 2003)
BCA2	Rab7 (degradation)	Breast	(Sakane et al. 2007)
	EGFR (stabilization)	Prostate, renal	(Burger et al. 2006)
C-Cbl	EGFR, Her2/neu, PDGFR	Glioblastoma	(Mukherjee et al. 2006)
	IGFR1 (degradation)		(Mizoguchi et al. 2004)
Bmi1/RING1A	histone H2A	Lymphoma, brain	(Jacobs and von Lohuizen 2002; Bruggeman et al. 2005; Liu et al. 2006)
PHD-domain E3s			
MEKK1	ERK1/2	Melanoma, pancreas	(Smalley 2003)
			(Li et al. 2005)
U-box E3s			
CHIP	Her2/neu (degradation)	Breast, ovarian	(Yarden 2001)
			(Hatakeyama and Nakayama 2003)

(e.g., their substrates and interacting partners). Interestingly, deubiquitinating enzymes, or E2s, were not overrepresented among cancer genes. Thus, E3 genes represent primary targets as cancer susceptibility genes for mutation screening and for the design of novel therapies (Beckmann et al. 2004, 2005).

HECT-Type E3 Ligases

The HECT family of E3 ligases carries a 350 amino acid residue domain homologous to the COOH-terminal domain of the prototype member of the family, E6AP (Scarafia et al. 2000). The conserved carboxyterminal region is involved in the formation of the ubiquitin thioester intermediate that is required for substrate ubiquitinylation. Twenty-eight HECT domain genes are known. Structurally, the HECT E3s are large proteins with multiple binding domains; in particular, they contain WW domains (the WW domain is a 35–40 amino acid repeat containing two tryptophan residues) that are thought to be critical for substrate recognition (Fig. 1a). The best-characterized HECT domain E3s are E6AP and members of the Nedd4-like E3 family. E6AP, WWP1, Nedd4, Smurf1, and Smurf2 have been linked with cancer (Chen and Matesic 2007; Scarafia et al. 2000) (Table 1).

E6AP

This HECT ligase was the first ligase found to degrade the tumor suppressor p53. E6AP requires the E6 protein of the papilloma viruses HPV types 16 or 18 for mediating binding of the E6-AP ubiquitin ligase to the p53 tumor suppressor protein (Brown and Pagano 1997; Pagano and Benmaamar 2003). Thus, the oncogenic activity of E6AP is important in cancers caused by HVP 16/18 infections such as cervical cancer and provides a mechanism for p53 degradation in the presence of E6. Because p53 is a sequence-specific DNA-binding transcription factor that responds to DNA damage and stress, and activates cell cycle arrest and apoptosis to prevent the propagation of damaged cells, its inactivation permits the accumulation of genetic damage, leading to transformation and tumorigenesis (Nalepa and Harper 2003; Burger and Seth 2004).

Nedd4-Like E3s

Nine Nedd4-like E3s exist. Nedd4-like E3s themselves are frequently regulated by phosphorylation, ubiquitination, translocation, and transcription in cancer. Because Nedd4-like E3s regulate ubiquitin-mediated trafficking, lysosomal or proteasomal degradation, and nuclear translocation of multiple proteins, they modulate important pathways involved in tumorigenesis such as TGFβ-, EGF-, IGF-, VEGF-, and TNFα-mediated signaling. Additionally, several Nedd4-like E3s directly regulate various cancer-related transcription factors from the Smad, p53, RUNX, and Jun families. Four of them, namely, Nedd4-1, Smurf1, Smurf2, and WWP1 have important function in cancer (Chen and Matesic 2007) (Table 1).

Nedd4-1

Nedd4-1 has recently been identified as the E3 ligase for the tumor suppressor protein PTEN (Wang et al. 2007). Nedd4 can ubiqutinylate PTEN in vitro and in vivo. Although NEDD4-1 alone was not oncogenic, its overexpression increased the efficiency of Ras-mediated transformation in *Trp53*-deficient mouse embryonic fibroblasts. Further analysis indicated that upregulation of NEDD4-1 can post-translationally suppress PTEN activity in human and mouse cancers. NEDD4-1 is a potential proto-oncogene that negatively regulates PTEN through ubiquitinylation (Baker 2007). Even though Nedd4-1 expression has not been widely studied in human tumors, the fact that PTEN is its substrate suggests essential roles of this E3 in tumors that can be associated with loss of PTEN expression such as prostate and endometrial cancers.

Smurf2

Smurf2 is a Nedd4-like E3 that was found to negatively regulate Smad signaling. Smad proteins serve as key signaling effectors of the TGF-β superfamily of growth factors. TGF-β signaling requires the action of Smad proteins in association with other coactivator and corepressor proteins to modulate target gene transcription. Smurf2 ligase displays preferences for Smad2 compared to other receptor-activated Smads (R-smads). Smurf2 ubiquitinylation of Smad2 reduces its ability to promote transcription and hence regulates the competence of a cell to respond to TGF-β signaling, a pathway that is particularly important in breast cancer (Zhang et al. 2001; Attisano and Wrana 2002; Subramaniam et al. 2003; Azmi and Seth 2005).

We have recently identified a small RING-H2-finger protein RNF11 that is associated with Smurf2 E3 ligase (Fig. 1a). RNF11 binding to Smurf2 in mammalian cells suggests that similar to Smads 2 and 3, it recruits targets for destruction by the Smurf2 E3 ligase. The most interesting RNF11/Smurf2 target that has been reported is AMSH (Li and Seth 2004; Azmi and Seth 2005).

AMSH (associated molecule with the SH3 domain of stam) is a deubiquitinating enzyme with functions at the endosome, which oppose the ubiquitin-dependent sorting of receptors to lysosomes. Specifically, AMSH is an important participant in multivesicular bodies pathway degradation of EGFR and enhances the degradation rate of EGF receptor (EGFR) following acute stimulation (McCullough et al. 2004; Ma et al. 2007).

Most interestingly, the Smurf2 E3 ligase and the RNF11 protein are both strongly expressed in head and neck (H&N) tumors, a tumor type also known for overexpression of EGFR, suggesting that the degradation of AMSH by Smurf2/RNF11

(Fig. 1a) might in part be responsible for high EGFR levels in H&N tumors (Ang et al. 2002; Fukuchi et al. 2002; Subramaniam et al. 2003).

Smurf1

Smurf1 also belongs to the Nedd4-like E3s and controls the local levels of the small guanosine triphosphatase RhoA. Thus, Smurf1 regulates cell polarity and protrusive activity. Smurf1 was found to be required to maintain the transformed morphology and motility of tumor cells (Wang et al. 2003a). RhoA degradation in tumor cells results in the downregulation of Rho kinase (ROCK) activity and myosin light chain 2 (MLC2) phosphorylation at the cell periphery. The localized inhibition of contractile forces is necessary for the formation of lamellipodia and for tumor cell motility (Sahai et al. 2006). A recent expression profiling of amplified genes in pancreatic adenocarcinomas identified Smurf1 as a potential novel oncogene in this tumor type (Loukopoulos et al. 2007) (Table 1).

WWP1

WWP1 is a Nedd4-like E3 ubiquitin ligase that contains four tandem WW domains and a HECT domain. WWP1 plays physiological roles in receptor trafficking and transcription. One of its substrate is the transcription factor KLF5. WWP1 mediates the ubiquitination and degradation of KLF5, which was identified as a candidate tumor suppressor gene in prostate and breast cancers (Chen et al. 2007a, b). WWP1 also negatively regulates the TGFβ tumor suppressor pathway by mediating the degradation Smad2, Smad4, and TGFβ receptor 1 (TβR1) proteins. The latter are implicated in breast cancer development and progression (Chen et al. 2007b) (Table 1).

Interestingly, RNF11 has been shown to interact with several HECT-type E3 ubiquitin ligases, including Nedd4, Smurf1 and Smurf2, as well as Cullin1, the core protein in the multisubunit SCF E3 ubiquitin ligase complex described later (Fig. 1a, c). Smad2 can also be a cofactor for multiple HECT E3 (e.g., Smurf2, WWP1). However, it appears as if the HECT ligase binding partner or the modifier/cofactor molecules differ within a certain context and might depend on the local environment. Thus, the co-factor molecule likely determines tissue and tumor type specificity of the ligase.

Adaptor-Type E3 Ligases

The adaptor-type E3 ligases comprise RING-finger, U-box, and PHD domain containing proteins (Fig. 1b–d). The RING-finger E3s can be monomeric (Fig. 1b) or are composed of multiple subunits (Fig. 1c) (Burger and Seth 2004). PHD domain and U-box-type E3 ligases have only recently been discovered. Their three-dimensional structure has a very high similarity to RING-finger E3 ligases, and little is know about their function. Most of the known and well characterized E3 ligases that play a role in cancer development and progression are RING-finger-type E3 ligases (Table 1).

RING-Finger E3 Ligases

RING E3s in contrast to HECT-type ligases mediate the direct transfer of E2 bound ubiquitin to substrates without a thioester bond formation (Fig. 1).

When RING-finger proteins were discovered, their only known role was the involvement in the dimerization of proteins. RING E3 ligases became subject of intense studies for their roles in cancer when evidence evolved of their crucial function in mediating transfer of ubiquitin to heterologous substrates (ubiquitination), with key roles in cell signaling and survival pathways, such as p27 or p53, leading to their degradation (Fig. 1b, c) (Joazeiro and Weissmann 2000; Burger et al. 2006). RING-fingers are classified as RING-HC-type (C3HC4) and RING-H2-type (C3H2C3) proteins (Fig. 2). The cysteines (C) and histidines (H) represent zinc-binding residues. The RING domain forms a distinct so-called cross-brace structure, in which metal–ligand pairs 1 and 3 coordinate to bind one zinc ion, and pairs 2 and 4 bind the second one (Fig. 2a). Based on the crystal structure of E2-E3 complexes such as UbcH7 (E2) and c-Cbl (E3), it could be shown that RING-fingers have E3 ligase activity and contribute to the catalysis of the reaction that brings together the E2 active site and the substrate's acceptor lysine(s) (Fig. 1) (Zheng et al. 2000; Kosarev et al. 2002).

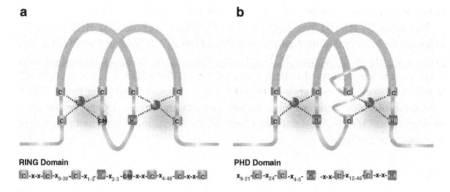

Fig. 2 RING and PHD domain E3 ligase cross-brace structure. (**a**) The RING-finger E3s are classified into RING-HC-type and RING-H2-type proteins. The only difference between these two RING proteins is in one amino acid that can either be a histidine (H) or a cysteine (C) as indicated by C/H in the cross-brace structure. (**b**) The PHD domain resembles closely the RING domain, but differences are in the alignment of histidines and cysteines, leading to slight differences in the cross-brace. Characteristics of the H- and C-rich RING and PHD domains are their zinc complexing ability. Two Zn2+ ions are complexed per cross-brace structure

Multisubunit Complexes

SCF: The small RING-finger protein Rbx1/Roc1 was among the first to be found associated with ubiquitination (Joazeiro and Weissmann 2000). It is an essential component of the multisubunit complex SCF (*S*kp1/*C*ullin-1/*F*-box protein) E3 ligase (Fig. 1c). SCF complexes are a prominent family of ubiquitin–protein ligases that ubiquitinate a broad range of substrates involved in cell cycle progression, signal transduction, and transcription (Otha and Fukuda 2004; Burger et al. 2006). Notably, SCF is the E3 ligase that regulates G1/S transition, because key proteins in this process such as p27 are substrates of the SCF (Fig. 1c). While the Roc1 RING-finger domain interacts with an E2 and provides E3 ligase activity, the F-box subunit provides the receptor for specific substrates (Zheng et al. 2002; Burger and Seth 2004). In primary breast cancers, p27 levels were found to be reduced and to predict poor prognosis, suggesting enhanced degradation of p27 by SCF (Catzavelos et al. 1997). The RING finger protein 11 (RNF11) can bind to Cullin1 and replace Roc1 within the SCF (Fig. 1c). In our laboratory, we found that RNF11 is overexpressed in invasive breast cancers, consistent with an enhanced activity of the SCF in this tumor type and hence degradation of SCF substrates such as p27 (Fig. 1c). Other substrates of the SCF are PTEN, Rb, and NFκB (Burger and Seth 2004; Azmi and Seth 2005).

VHL (von Hippel-Lindau): E3 ligase complex is another multisubunit E3 with a role in cancer. Mutations in the VHL gene are involved in the family cancer syndrome for which it is named and in the development of sporadic renal cell cancer (RCC) (Mani and Gelman 2005). VHL is in a complex with elongin B and elongin C, Cullin2, and Rbx1/Roc1. The VHL ubiquitin ligase complex targets members of the hypoxia-inducible transcription factor family (HIF) for degradation under normoxic conditions (Tyers and Rotappel 1999). Hypoxia-inducible factors mediate a physiologic response to hypoxia by activating the expression of genes that promote angiogenesis, such as vascular endothelial growth factor (VEGF). As a component of a multiprotein ubiquitin E3 ligase complex, VHL interacts directly with the oxygen-dependent destruction domain of the hypoxia-inducible factors. Mutations in VHL prevent the degradation of HIFα-subunits under normoxic conditions and predispose to the formation of hypervascular lesions and renal tumors. HIF upregulation in renal cell carcinoma is correlated with enhanced expression of VEGF and other hypoxia-inducible genes (Mani and Gelmann 2005).

APC/C: Anaphase-promoting complex/cyclosome is a multisubunit E3 ligase required for ubiquitin-dependent proteolysis of cell-cycle-regulatory proteins, including mitotic cyclins and securin/Pds1. APC/C is composed of 13 core subunits, most of which are conserved in all eukaryotes. Regulatory phosphorylations and coactivator proteins (Cdc20/fizzy or Cdh1/Hct1/fizzy-related) activate APC/C to ubiquitinylate substrates containing destruction (D) and/or KEN boxes in a cell-cycle-dependent manner (Passmore and Barford 2005). Cdh1-APC targets the F box protein Skp2^{SCF}, which is a part of the SCF E3 ligase, for degradation in response to TGF-β, resulting in an increased half-life of p27 (Fig. 1c) (Liu et al. 2007). Accumulation of p27 induced by TGF-β contributes to the inhibition of Cdk2/cyclin E, a major driver for the G_1/S transition, and leads to cell cycle arrest. Impairment of APC is frequently detected in various human cancers, including colon (Wang et al. 2003b) and breast cancer (Park et al. 2005).

Monomeric RING E3s

Monomeric RING-finger E3 ligases unlike those participating in the SCF complex have a substrate binding domain and are capable of substrate ubiquitination as well as of autoubiquitination (Fig. 1b). The latter is a hallmark of these E3s. The mono-meric RING-finger family of E3s include but are not limited to Mdm2, Efp, c-Cbl, BRCA1/BARD1, COP1, Bmi-1, and BCA2 (Fig. 1b, Table 1) (Joazeiro et al. 1999; Urano et al. 2002; Dornan et al. 2004; Feki et al. 2005; Burger et al. 2005; Brooks and Gu 2006; Buchwald et al. 2006). COP1, Mdm2, and BRCA1/BARD1 have a common substrate, namely p53 (Dornan et al. 2004; Feki et al. 2005; Brooks and Gu 2006).

COP1 has been found to be significantly overexpressed in 81% of breast carci-nomas that were profiled for COP1 expression and concomitantly p53 was decreased (Dornan et al. 2004).

Germline mutations in the RING finger of *BRCA1* predispose women to the early onset of breast tumors. They account for approximately 5% of breast cancers. BRCA1 ubiquitin ligase activity is enhanced when it is dimerized with BRCA1-associated RING domain (BARD1) and has been implicated in dysregulation of p53-associated DNA damage response (Yoshikawa et al. 2000; Dong et al. 2003). While COP1 and BRCA1 seem to regulate p53 stability in breast cancer, *Mdm2* appears to be more associated with poor outcome of other tumor types, for example, sarcoma and glioma (Onel and Cordon-Cardo 2004).

Other substrates of RING-type E3 ligases are 14-3-3σ and activated tyrosine kinase receptors. 14-3-3σ is a p53-regulated inhibitor of G2/M progression and was shown to have tumor suppressor function in breast cancer. 14-3-3σ is degraded by the estrogen-responsive finger protein (*Efp*) (Urano et al. 2002; Horie et al. 2003). Epidermal growth factor receptor 1 (EGFR) and 2 (Her2/neu) are substrates of *c-Cbl*. c-Cbl recognizes phosphorylated tyrosines in receptor tyrosine kinases through its SH2 domain and negatively regulates signaling by facilitating receptor ubiquitination (Levkowitz et al. 2000; Duan et al. 2003; Mukherjee et al. 2006).

The Breast Cancer Associated gene 2 *(BCA2)* (synonymous to T3A12/ZNF364/ Rabring7) was discovered by us in an effort to identify novel breast cancer associated genes by subtractive hybridization cloning (Burger et al. 1998). BCA2 has a RING-H2 consensus sequence and complexes two zinc ions (Figs. 1b and 2a). We have previously reported that BCA2 has E3 ligase activity, and in that we showed that it is capable of autoubiquitination. The latter is dependent on the RING-domain as demonstrated by the loss of E3 ligase activity in mutant RING-BCA2 proteins (Burger et al. 2005). We also found that BCA2 is expressed in invasive breast cancers and that its expression is associated with positive estrogen receptor status and patient outcome. Other tumor types that were found to express BCA2 are prostate cancers and papillary renal cell carcinomas. BCA2 has a distinct nuclear and cytoplasmic expression pattern in cancer cells (Burger et al. 2006). In the cytoplasm, BCA2 colocalizes with small G protein Rab7. We and others have shown that BCA2, also termed Rabring7, is a target protein of Rab7. Rab7 and other Rab family members regulate endocytic trafficking and hence growth factor receptor degradation in the lysosomes. It was found that overexpression of BCA2 significantly inhibits Rab7-mediated epidermal growth factor (EGF) trafficking into the late endosome and subsequently its degradation in the lysosomes (Mizuno et al. 2003; Burger et al. 2006; Sakane et al. 2007).

Interestingly, BCA2 had also a 14-3-3 binding domain that contains an AKT phosphorylation site. AKT mutant BCA2 protein does not bind to Rab7, indicating that phosphorylation by AKT might be essential for the BCA2:Rab7 interaction and that tumors with an activated AKT pathway could upregulate growth factor receptor signaling by facilitating their recycling and preventing sorting into the lysosomes when Rab7 is recruited to BCA2 (Connor et al. 2005; Burger et al. 2006).

Thus, the interaction between the E3 ligase BCA2 and Rab7 suggests a major role of these proteins in the cross talk between the EGFR and AKT signaling pathways and further supports the major role of kinases and ligases in regulating cancer cell growth.

c-Cbl is an E3-ligase with a RING-finger and a SH2 domain. It recognizes phosphorylated tyrosines on receptor tyrosine kinases through its SH2 domain and negatively regulates signaling by facilitating receptor ubiquitination. C-Cbl recognizes tyrosine-phosphorylated receptors (RTK) and E2 ubiquitin conjugating enzymes, and brings them together to enforce their degradation. First c-Cbl and the multiadapter protein growth factor receptor-binding protein 2 (Grb2) bind to the activated RTK and provide further docking sites for cbl-interacting protein of 85 k (CIN85) or EGFR pathway substrate-15 (Eps15). These adapter proteins, in turn, recruit disabled2 (DAB2) and clathrin adaptor protein-2 (AP2) as well as endophilin and dynamin-2 that initiate endocytosis by facilitating budding. Following internalization, the vesicles fuse with Rab5-positive early endosomes. From early endosomes, receptors may recycle back to the plasma membrane or be sorted to late endosomes for lysosomal and/or proteasomal degradation (d'Azzo et al. 2005; Mukherjee et al. 2006).

EGFR, insulin receptor (IR); insulin-like growth factor receptor 1 (IGFR1), platelet-derived growth factor receptor (PDGF-R), and Her2/neu have been described as c-Cbl substrates. All of which are critical RTKs in cancer development and progression in a variety of tumor types that include breast, brain, soft tissue sarcoma, and head and neck cancers (Mizoguchi et al. 2004; Mukherjee et al. 2006).

Bmi1 contains an N-terminal RING-domain and is a member of the polycomb group (PcG) proteins that adjust chromatin conformation and mediate target gene repression. Bmi1 forms a heterodimer with RING1A, similar to the BRCA1/BARD1 complex, and then assembles with other PcG proteins into the maintenance complex Polycomb repressive complex 1 (PRC1), which is involved in developmentally regulated and tissue-specific transcriptional silencing. The PRC1 ubiquitin E3 ligase monoubiquitinates histone H2A at lysine 119.

Within the PRC1 complex, the Bmi1 protein has a unique role that is critical for stem cell maintenance and cancer formation (Buchwald et al. 2006; Chumsri et al. 2007). It has a strong dose-dependent phenotype, where overexpression can lead to

B- and T-cell lymphomas and where partial reduction of Bmi1 leads to significant reduction in lymphoma formation (Jacobs and van Lohuizen 2002) and brain tumour formation (Bruggeman et al. 2005). Recent work has shown that Bmi1 is important in regulating self-renewal of normal and malignant human mammary stem cells (Liu et al. 2006).

PHD and U-Box E3 Ligases

PHD Domain E3 Ligases

PHD proteins closely resemble the RING-finger domain (Figs. 1d and 2). They are also a specialized form of zinc fingers. The cross-brace structure of the PHD domain (Fig. 2b) is very similar to that of the RING domain (Fig. 2a), differing just in the spacing and distribution of the cysteine and histidine residues. Only very few human PHD-type E3 ligases have so far been described (Table 1) (Lu et al. 2002; Burger and Seth 2004; Lu and Hunter 2008).

MEKK1: The most prominent PHD E3 is the MAP kinase kinase kinase MEKK1 (Fig. 1d). The dual function of MEKK1 as a protein kinase and as a ubiquitin protein ligase was first described by Hunter and co-workers, who found that MEKK1 exhibits E3 ligase activity towards ERK1/2. The E3 activity could be abolished by mutations of conserved cysteines in the PHD domain (Lu et al. 2002; Lu and Hunter 2008). ERK1/2 is widely expressed in tumor cells and provides a protective effect against apoptosis, while constitutive MEKK1 activation is sufficient to cause apoptosis and to downregulate ERK1/2 survival signals through UP-S degradation pathways (Johnson and Lapadat 2002). Tumors highly dependent on the mitogen-activated protein kinases (MAPKs) ERK1/2 are malignant melanomas and pancreatic cancers (Table 1) (Smalley 2003; Li et al. 2005).

U-Box E3 Ligases

U-box proteins were identified as a distinct class of adaptor-type E3 ligases because of their similarity with the three-dimensional structure of the RING ligases. The conserved zinc-binding sites supporting the cross-brace arrangement in RING-finger domains are replaced by hydrogen-bonding networks in the U-box (Hatakeyama et al. 2000; Ohi et al. 2003; Hatakeyama and Nakayama 2003).

C terminus of Hsp70 interacting protein (*CHIP*) is the best characterized U-box ligase. It contains two tetratricopeptide repeat (TPR) domains that facilitate interaction with the carboxyl terminus of the molecular chaperones Hsp70 or Hsp90 (Hatakeyama and Nakayama 2003). While the U-box domain executes the E3 ligase activity, the TPR motif associates with heat shock proteins, and the CHIP/Hsp complex then acts as a protein quality control ubiquitin ligase that is able to selectively target abnormally folded proteins for destruction in the proteasome (Murata et al. 2003). CHIP plays an important role in cancer, because it mediates the degradation of "abnormal proteins" such as Her2/neu that are involved in the development and/or aggressiveness of several tumor types, including breast and ovarian cancers. Because of Her2/Neu persistence and resistance to degradation and because of its propensity for dimerization with other Her family members such as Her1 and Her3, its overexpression is associated with poor disease prognosis. Interestingly, the CHIP/Her2 interaction can be enhanced by Hsp90 inhibitors such as geldanamycin (GA) and 17-AAG (17-allylamino-geldanamycin). 17-AAG is currently in clinical trials for the treatment of cancer (Yarden 2001; Xu et al. 2002; Burger 2007).

Conclusions and Future Perspectives

More and more evidences emerge that ubiquitin ligases impact the development of cancer by regulating receptor signaling initiated by growth factors, cytokines, or hormones, as well as cell survival and cell death decisions.

Results of intense cancer gene discovery efforts over the past decade revealed that ubiquitin E3 ligases have a distinct role in cancer (Table 1). In particular, breast cancers appear to have an aberrant expression of several E3 ligases and their substrates.

Four different scenarios exist that link E3 ligases to cancer development and progression: (a) E3 ligases that stabilize tumor suppressor proteins such as p27 are impaired (APC/C); (b) E3 ligases that degrade tumor suppressor proteins such p53 (Mdm2) or 14-3-3σ (Efp) are upregulated; (c) E3 ligases that degrade oncogenic receptor tyrosine kinases (c-Cbl, CHIP) are deregulated; and (d) E3 ligases that stabilize oncogenic receptor tyrosine kinases (BCA2) are overexpressed.

The targeting of the protein–protein interactions between the E3s and their prominent substrates offer specific and very attractive approaches to tumor therapy, and it is highly likely that tumor-type-specific E3 ligase signatures will emerge as our knowledge base increases.

Currently, the best known function of E3 ligases is polyubiquitination of substrates, targeting them for destruction in the proteasome. However, only polyubiquitinated proteins with at least four ubiquitin molecules linked via ubiquitin lysine residue 48 are targeted to the proteasome and undergo degradation. Ubiquitin itself has seven conserved lysine residues and all can potentially form isopetidyl linkages on the carboxyl terminus of another ubiquitin; these ubiquitin modifications are implicated in regulating DNA repair, transcription, and vesicular transport (Weissman 2001; Arndt and Winston 2005; Mukherjee et al. 2006). The mechanisms that allow E3 ligases to perform such different linkages are poorly understood. The understanding of these mechanisms is a future challenge, and they also offer the opportunity of great progress in delineating E3 physiological functions and cancer cell biology. Moreover, other ubiquitination factors, termed E4, that seem to be involved in multiubiquitin chain assembly (Koegl et al. 1999) and deubiquitinating enzymes (Wilkinson 1997) need to be studied in the context of E3 function. Of further importance, E3 ligases can not only perform ubiquitination processes, but also attach ubiquitin-like molecules such as SUMO or NEDD8 to substrates, indicating that ubiquitination is more complex than believed (Sentis et al. 2005; Schmidt and Dikic 2006).

Hence, interference with ubiquitin ligases can be exploited not only to directly stabilize desirable and/or enforce the degradation of unwanted proteins, but also to modulate other essential processes such as transcription and posttranslational protein modification. The modulation of E3 ligases promises specificity and effectiveness that might exceed the success of kinase inhibitors in anticancer therapy.

Acknowledgements We thank Mrs. Alicia Castillo for her assistance with the artwork. This work was funded by R01CA127258-01(AMB) and the Canadian Breast Cancer Research Alliance special program grant on metastasis (AKS).

References

Ang, K.K., Berkey, B.A., Tu, X., Zhang, H.Z., Katz, R., Hammond, E.H., Fu, K.K., and Milas, L. 2002. Impact of epidermal growth factor receptor expression on survival and pattern of relapse in patients with advanced head and neck carcinoma. Cancer Res 62:7350–7356.

Arndt, K., and Winston, F. 2005. An unexpected role for ubiquitylation of a transcriptional activator. Cell 120:733–737.

Attisano, L., and Wrana, J.L. 2002. Signal transduction by the TGF-beta superfamily. Science 296:1646–1647.

Azmi, P., and Seth, A. 2005. RNF11 is a multifunctional modulator of growth factor receptor signalling and transcriptional regulation. Eur J Cancer 41:2549–2560.

Baker, S.J. 2007. PTEN enters the nuclear age. Cell 128:25–28.

Bohley, P. 1995. The fates of proteins in cells. Naturwissenschaften 82:544–550.

Beckmann, J.S., Maurer, F., Delorenz, M., and Falquet, L. 2004. Ubiquitin ligases as cancer genes. Nat Rev Cancer 4:doi:10.1038/nrc1299-c1.

Beckmann, J.S., Maurer, F., Delorenz, M., and Falquet, L. 2005. On ubiquitin ligases and cancer. Hum Mutat 25:507–512.

Brooks, C.L., and Gu, W. 2006. p53 ubiquitination: Mdm2 and beyond. Mol Cell 21:307–315.

Brown, J.P., and Pagano, M. 1997. Mechanism of p53 degradation. Biochim Biophys Acta 1332:1–6

Bruggeman, S.W., Valk-Lingbeek, M.E., van der Stoop, P.P., Jacobs, J.J., Kieboom, K., Tanger, E., Hulsman, D., Leung, C., Arsenijevic, Y., Marino, S., and van Lohuizen, M. 2005. Ink4a and Arf differentially affect cell proliferation and neural stem cell self-renewal in Bmi1-deficient mice. Genes Dev 19:1438–1443.

Buchwald, G., van der Stoop, P., Weichenrieder, O., Perrakis, A., van Lohuizen, M., and Sixma, T.K. 2006. Structure and E3-ligase activity of the ring-ring complex of Polycomb proteins Bmi1 and Ring1b. EMBO J 25:2465–2474.

Burger, A., Li, H., Zhang, X.K., Venanzoni, M., Vournakis, J., Papas, T., and Seth, A. 1998. Breast cancer genome anatomy: correlation of morphological changes in breast carcinomas with expression of the novel gene product Di12. Oncogene 16:327–333.

Burger, A.M., and Seth, A.K. 2004. The ubiquitin mediated protein degradation pathway in cancer: therapeutic implications. Eur J Cancer 40:2217–2229.

Burger, A.M., Gao, Y., Amemiya, Y., Kahn, H.J., Kitching, R., Yang, Y., Sun, P., Narod, S.A., Hanna, W.M., and Seth, A.K. 2005. A novel RING-type ubiquitin ligase breast cancer-associated gene 2 correlates with outcome in invasive breast cancer. Cancer Res 65:10401–10412.

Burger, A., Amemiya, Y., Kitching, R., and Seth, A.K. 2006. Novel RING E3 ubiquitin ligases in breast cancer. Neoplasia 8:689–695.

Burger, A.M. 2007. Highlights in experimental therapeutics. Cancer Lett 245:11–21.

Catzavelos, C., Bhattacharya, N., Ung, Y.C., Wilson, J.A., Roncari, L., Sandhu, C., Shaw, P., Yeger, H., Morava-Protzner, I., Kapusta, L., Franssen, E., Pritchard, K.I., and Slingerland, J.M. 1997. Decreased levels of the cell-cycle inhibitor p27Kip1 protein: prognostic implications in primary breast cancer. Nat Med 3:227–230.

Ciechanover, A. 2005. Proteolysis: from the lysosome to the ubiquitin and the proteasome. Nat Rev Mol Cell Biol 6:79–86.

Chen, C., Sun, X., Guo, P., Dong, X.Y., Sethi, P., Zhou, W., Zhou, A., Petros, J., Frierson Jr, H.F., Vessella, R.L., Atfi, A., Dong, J.-T. 2007a. Ubiquitin E3 ligase WWP1 as an oncogenic factor in human prostate cancer. Oncogene 26:2386–2394.

Chen, C., Zhou, Z., Ross, J.S., Zhou, W., and Dong, J.T. 2007b. The amplified WWP1 gene is a potential molecular target in breast cancer. Int J Cancer 121:80–87.

Chen, C., and Matesic, L.E. 2007. The Nedd4-like family of E3 ubiquitin ligases and cancer. Cancer Metastasis Rev 26:587–604.

Chumsri, S., Matsui, W., and Burger, A.M. 2007. Leukemic stem cell pathways – therapeutic implications. Clin Cancer Res 13:6549–6554.

Coscoy, L., and Ganem, D. 2003. PHD domains and E3 ubiquitin ligases: viruses make the connection. Trends Cell Biol 13:7–12.

Connor, M.K., Azmi, P.B., Subramaniam, V., Li, H., and Seth, A. 2005. Molecular characterization of Ring-finger protein 11. Mol Cancer Res 3:453–461.

d'Azzo, A., Bongiovanni, A., and Nastasi, T. 2005. E3 ubiquitin ligases as regulators of membrane protein trafficking and degradation. Traffic 6:429–441.

Dong, Y., Hakimi, M.A., Chen, X., Kumaraswamy, E., Cooch, N.S., Godwin, A.K., and Shiekhattar, R. 2003. Regulation of BRCC, a holoenzyme complex containing BRCA1 and BRCA2, by a signalosome-like subunit and its role in DNA repair. Mol Cell 12:1087–1099.

Dornan, D., Bheddah, S., Newton, K., Ince, W., Frantz, G.D., Dowd, P., Koeppen, H., Dixit, V.M., and French, D.M. 2004. COP1, the negative regulator of p53, is overexpressed in breast and ovarian adenocarcinomas. Cancer Res 64:7226–7230.

Duan, L., Miura, Y., Dimri, M., Majumder, B., Dodge, I.L., Reddi, A.L., Ghosh, A., Fernandes, N., Zhou, P., Mullane-Robinson, K., Rao, N., Donoghue, S., Rogers, R.A., Bowtell, D., Naramura, M., Gu, H., Band, V., and Band, H. 2003. Cbl-mediated ubiquitinylation is required for lyso-somal sorting of epidermal growth factor receptor but is dispensable for endocytosis. J Biol Chem 278:28950–28960.

Feki, A., Jefford, C.E., Berardi, P., Wu, J.Y., Cartier, L., Krause, K.H., and Irminger-Finger, I. 2005. BARD1 induces apoptosis by catalysing phosphorylation of p53 by DNA-damage response kinase. Oncogene 24:3726–3736.

Fukuchi, M., Fukai, Y., Masuda, N., Miyazaki, T., Nakajima, M., Sohda, M., Manda, R., Tsukada, K., Kato, H., and Kuwano, H. 2002. High-level expression of the Smad ubiquitin ligase Smurf2 correlates with poor prognosis in patients with esophageal squamous cell carcinoma. Cancer Res 62:7162–7165.

Futreal, P.A., Coin, L., Marshall, M., Down, T., Hubbard, T., Wooster, R., Rahman, N., and Stratton, M.R. 2004. A census of human cancer genes. Nat Rev Cancer 4:177–183.

Glickman, M.H., and Ciechanover, A. 2002. The ubiquitin-proteasome proteolytic pathway: destruction for the sake of construction. Physiol Rev 82:373–428.

Goldberg, A.L., Stein, R., and Adams, J. 1995. New insights into proteasome function: from archaebacteria to drug development. Chem Biol 2:503–508.

Handley-Gaerhart, P.M., Stephen, A.G., Trausch-Azar, J., Ciechanover, A., and Schwartz, A.L. 1994. Human ubiquitin activating enzyme E1, indication of potential nuclear and cytoplasmic subpopulations using epitope-tagged cDNA constructs. J Biol Chem 269:33171–33178.

Hatakeyama, S., Yada, M., Matsumoto, M., Ishida, N., and Nakayama, K.I. 2000. U-box proteins as a new family of ubiquitin-protein ligases. J Biol Chem 276:33111–33120.

Hatakeyama, S., and Nakayama, K.I. 2003. U-box proteins as a new family of ubiquitin ligases. Biochem Biophys Res Comm 302:625–645.

Hershko, A., Ciechanover, A., and Varshafsky, A. 2000. The ubiquitin system. Nat Med 10:1073–1074.

Hershko, A., Ciechanover, A., Heller, H., Haas, A.L., and Rose, I.A. 1980. Proposed role of ATP in protein breakdown: conjugation of protein with multiple chains of the polypeptide of ATP-dependent proteolysis. Proc Natl Acad Sci 77:1783–1786.

Hicke, L. 2001. Protein regulation by monoubiquitin. Nat Rev 2:195–201.

Horie, K., Urano, T., Ikeda, K., and Inoue, S. 2003. Estorgen-responsive RING-finger protein controls breast cancer growth. J Steroid Biochem Mol Biol 85:101–104.

Jacobs, J.J., and van Lohuizen, M. 2002. Polycomb repression: from cellular memory to cellular proliferation and cancer. Biochim Biophys Acta 1602:151–161.

Joazeiro, C.A., Wing, S.S., Huang, H., Leverson, J.D., Hunter, T., and Liu, Y.C. 1999. The tyrosine kinase negative regulator cCbl as a RING-type E2-dependent ubiquitin-protein ligase. Science 286:309–312.

Joazeiro, C.A.P., and Weissmann, A.M. 2000. RING finger proteins: mediators of ubiquitin ligase activity. Cell 102:549–552.

Johnson, G.L., and Lapadat, R. 2002. Mitogen-activated protein kinase pathways mediated by ERK, JNK, and p38 protein kinases. Science 298:1911–1912.

Koegl, M., Hoppe, T., Schlenker, S., Ulrich, H.D., Mayer, T.U., and Jentsch, S. 1999. A novel ubiquitination factor, E4, is involved in multiubiquitin chain assembly. Cell 96:635–644.

Kosarev, P., Mayer, K.F.X., and Hardtke, C.S. 2002. Evaluation and classification of RING-finger domains encoded by the Arabidopsis genome. Genome Biol 3:0016.1–0016.12.

Lecker S.H., Goldberg , A.L., and Mitch, W.E. 2006. Protein degradation by the ubiquitin-proteasome pathway in normal and disease states. J Am Soc Nephrol 17:1807–1819.

Levkowitz, G., Oved, S., Klapper, L.N., Harari, D., Lavi, S., Sela, M., and Yarden, Y. 2000. c-Cbl is a suppressor of the neu oncogene. J Biol Chem 275:35532–35539.

Li, H.X., and Seth, A. 2004. An RNF11:Smurf2 complex mediates ubiquitination of the AMSH protein. Oncogene 23:1801–1808.

Li, M., Becnel, L.S., Li, W., Fisher, W.E., Chen, C., and Yao, Q. 2005. Signal transduction in human pancreatic cancer: roles of transforming growth factor a, somatostatin receptors, and other signal intermediates. Arch Immunol Ther Exp 53:381–387.

Liu, S., Dontu, G., Mantle, I.D., Patel, S., Ahn, N.S., Jackson, K.W., Suri, P., and Wicha, M.S. 2006. Hedgehog signaling and Bmi-1 regulate self-renewal of normal and malignant human mammary stem cells. Cancer Res 66:6063–6071.

Liu , W., Wu, G., Li, W., Lobur, D., and Wan, Y. 2007. Cdh1-anaphase-promoting complex targets Skp2 for destruction in transforming growth actor ß-induced growth inhibition. Mol Cell Biol 27:2967–2979.

Loukopoulos, P., Shibata, T., Katoh, H., Kokubu, A., Sakamoto, M., Yamazaki, K., Kosuge, T., Kanai, Y., Hosoda, F., Imoto, I., Ohki, M., Inazawa, J., and Hirohashi, S. 2007. Genome-wide array-based comparative genomic hybridization analysis of pancreatic adenocarcinoma: identification of genetic indicators that predict patient outcome. Cancer Sci 98:392–400.

Lu, Z., Xu, S., Joazeiro, C., Cobb, M.H., and Hunter, T. 2002. The PHD domain of MEKK1 acts as an E3 Ubiquitin Ligase and mediates ubiquitination ad degradation of ERK1/2. Mol Cell 9:945–956.

Lu, Z., and Hunter, T. 2008. MEKK1: dual function as a protein kinase and a ubiquitin protein ligase. 2008. In: Protein Degradation, vol. 2, Mayer , J., Ciechanover, A.J., and Rechsteiner, M. (eds.) Wiley, Germany, ISBN: 9783527620210.

Ma, Y.M., Boucrot, E., Villén, J., Affar el, B., Gygi, S.P., Göttlinger, H.G., and Kirchhausen, T. 2007. Targeting of AMSH to endosomes is required for epidermal growth factor receptor degradation. J Biol Chem 282:9805–9812.

Mani, A., and Gelmann, E.P. 2005. The ubiquitin-proteasome pathway and its role in cancer.J Clin Oncol 23:4776–4789.

McCullough, J., Clague, M.J., and Urbé, S. 2004. AMSH is an endosome-associated ubiquitin isopeptidase. J Cell Biol 166:487–492.

Mizoguchi, M., Nutt, C.L., and Louis, D.N. 2004. Mutation analysis of CBL-C and SPRED3 on 19q in human glioblastoma. Neurogenetics 5:81–82.

Mizuno, K., Kitamura, A., and Sasaki, T. 2003. Rabring7, a novel Rab7 target protein with a RING finger motif. Mol Biol Cell 14:3741–3752.

Mukherjee, S., Tessema, M., and Wandinger-Ness, A. 2006. Vesicular trafficking of tyrosine kinase receptors and associated proteins in the regulation of signaling and vascular function. Circulation Res 98:743–756.

Murata, S., Chiba, T., and Tanaka, K. 2003. CHIP: a quality-control E3 liagse collaborating with molecular chaperones. Int J Biochem Cell Biol 35:572–578.

Nalepa, G., and Harper, W. 2003. Therapeutic anti-cancer targets upstream of the proteasome. Cancer Treat Rev 29:49–57.

Ohi, M.D., Vander Kooi, C.W., Rosenberg, J.A., Chazin, W.J., and Gould, K.L. 2003. Structural insights into the U-box, a domain associated with multi-ubiquitination. Nat Struct Biol 10:250–255.

Ohta, T., and Fukuda, M. 2004. Ubiquitin and breast cancer. Oncogene 23:2079–2088.

Onel, K., and Cordon-Cardo, C. 2004. MDM2 and prognosis. Mol Cancer Res 2:1–8.

Pagano, M., and Benmaamar, R. 2003. When protein destruction runs amok, malignancy is on the loose. Cancer Cell 4:251–256.

Park, K. H., Choi, S.E., Eom, M., and Kang, Y. 2005. Downregulation of the anaphase-promoting complex (APC)7 in invasive ductal carcinomas of the breast and its clinicopathologic relationships. Breast Cancer Res 7:R238–R247.

Passmore, L.A., and Barford, D. 2005.Coactivator functions in a stoichiometric complex with anaphase-promoting complex/cyclosome to mediate substrate recognition. EMBO Rep 6:873–878.

Pray, T.R., Parlati, F., Huang, J., Wong, B.R., Payan, D.G., Bennett, M.K., Issakani, S.D., Molineaux, S., and Demo, S.D. 2002. Cell cycle regulatory E3 ubiquitin ligases as anticancer targets. Drug Resist Updates 5:249–258.

Sahai, E., Garcia-Medina, R., Pouysségur, J., and Vial, E. 2006. Smurf1 regulates tumor cell plasticity and motility through degradation of RhoA leading to localized inhibition of contractility. Cell Biol 176:35–42.

Sakane, A., Hatakeyama, S., and Sasaki, T. 2007. Involvement of Rabring7 in EGF receptor degradation as an E3 ligase. Biochem Biophys Res Comm 357:1058–1064.

Scarafia, L.E., Winter, A., and Swinney, D.C. 2000. Quantitative expression analysis of the cellular specificity of HECT-domain ubiquitin E3 ligases. Physiol Genomics 4:147–153.

Schmidt, J.J., and Dikic, I. 2006. Ubiquitin and NEDD8: brothers in arms. Sci STKE pe50:1–2.

Sentis, S., Le Romancer, M., Bianchin, C., Rostan, M.C., and Corbo, L. 2005. Sumoylation of the estrogen receptor αhinge region by SUMO-E3 ligases PIAS1 and PIAS3 regulates ERα transcriptional activity. Mol Endocrinol 19:2671–2684; doi:10.1210/me.2005–0042.

Smalley, K.S. 2003. A pivotal role for ERK in the oncogenic behaviour of malignant melanoma? Int J Cancer 104:527–532.

Subramaniam, V., Lubovitz, J., Li, H.X., Burger, A., Kitching, R., and Seth, A. 2003. The RING-H2 protein RNF11 is overexpressed in breast cancer and is a target of Smurf2 E3 ligase. Br J Cancer 89:1538–1544.

Tyers , M., and Rottapel , R. 1999. VHL: a very hip ligase. Proc Natl Acad Sci 96:12230–12232.

Urano, T., Saito, T., Tsukui, T., Fujita, M., Hosoi, T., Muramatsu, M., Ouchi, Y., and Inoue S. 2002. Efp targets 14.3.3σ for proteolysis and promotes breast tumour growth. Nature 417:871–875.

Wang, H.R., Zhang, Y., Ozdamar, B., Ogunjimi, A.A., Alexandrova, E., Thomsen, G.H., and Wrana, J.L. 2003a. Regulation of cell polarity and protrusion formation by targeting RhoA for degradation. Science 302:1775–1779.

Wang, Q., Moyret-Lalle, C., Couzon, F., Surbiguet-Clippe, C., Saurin, J.C., Lorca, T., Navarro, C., and Puisieux, A. 2003b. Alterations of anaphase-promoting complex genes in human colon cancer cells. Oncogene 22:1486–1490

Wang, X., Trotman, L.C., Koppie, T., Alimonti, A., Chen, Z., Gao, Z., Wang, J., Erdjument-Bromage, H., Tempst, P., Cordon-Cardo, C., Pandolfi, P.P., Jiang, X. 2007. NEDD4–1 is a proto-oncogenic ubiquitin ligase for PTEN. Cell 128:129–139.

Weissman, A.M. 2001. Themes and variations on ubiquitylation. Nat Rev 2:169–178.

Wilkinson, K.D. 1997. Regulation of ubiquitin-dependent processes by deubiquitinating enzymes. FASEB J 11:1245–1256.

Xu, W., Marcu, M., Yuan, X., Minnaugh, E., Patterson, C., and Neckers, L. 2002. Chaperone-dependent E3 ubiquitin ligase CHIP mediates a degradative pathway for c-ErbB2/Neu. Proc Natl Acad Sci USA 99:12847–12852.

Yang, Y., Li, C.C.H., and Weissman, A.M. 2004. Regulating the p53 system through ubiquitination. Oncogene 23:2096–2106.

Yarden, Y. 2001. Biology of Her2 and its importance in breast cancer. Oncology 61(Suppl 2):1–13

Yoshikawa, K., Ogawa, T., Baer, R., Hemmi, H., Honda, K., Yamauchi, A., Inamoto, T., Ko, K., Yazumi, S., Motoda, H., Kodama, H., Noguchi, S., Gazdar, A.F., Yamaoka, Y., and Takahashi, R. 2000. Abnormal expression of BRCA1 and BRCA1-interactive DNA-repair proteins in breast carcinomas. Int J Cancer 88:28–36.

Zhang, Y., Chang, C., Gehling, D.J., Hemmati-Brivanlou, A., and Derynck R. 2001. Regulation of Smad degradation and activity by Smurf2, an E3 ubiquitin ligase. Proc Natl Acad Sci USA 98:974–979.

Zheng, N., Wang, P., Jeffrey, P.D., and Pavletich, N.P. 2000. Structure of a c-Cbl-UbcH7 complex: RING domain function in ubiquitin-protein ligases. Cell 102:533–539.

Zheng, N., Schulman, B.A., Song, L., Miller, J.J., Jeffrey, P.D., Wang, P., Chu, C., Koepp, D.M., Elledge, S.J., Pagano, M., Conaway, R.C., Conaway, J.W., Harper, J.W., and Pavletich, N.P. 2002. Structure of the Cul1-Rbx1-Skp1-FboxSkp2 SCF ubiquitin ligase complex. Nature 416:703–708.

The 26S Proteasome as a Therapeutic Target in Cancer: Beyond Protease Inhibitors?

Rati Verma

Abstract The 26S proteasome is a multi-subunit complex that has a barrel-shaped peptidase core (CP) whose proteolytic activity is sequestered from the cellular milieu by two regulatory particles (RP) that are docked on either end. The RP confers ubiquitin and ATP dependence to the proteolytic process. The CP can also be regulated by the REGγ complex, and can, in some instances, catalyze the degradation of proteins independent of ATP and ubiquitin. The peptidase activity of the 20S CP has been validated as a therapeutic target for cancers such as multiple myeloma by the development of inhibitors such as VELCAID. Since the peptidase activity is so central to 26S proteasome function, inhibitors of this class are generally toxic. Alternative therapeutic targets within the 26S proteasome can be explored, given the multistep nature of regulated degradation. Specifically, the recognition, and deubiquitination, of the polyubiquitin chain, and the unfolding and gating steps can be targeted.

Keywords 26S proteasome • 26S proteasome inhibitors • Ubiquitin chain receptors • Unfoldases • Gating • Deubiquitination

The 26S proteasome catalyzes the ATP- and ubiquitin (Ub)-dependent proteolysis of regulatory proteins thereby committing the cell to irreversible transitions of the cell cycle or cell fate. The proteasome also eliminates un-folded and unassembled proteins (reviewed in Pickart and Cohen 2004). Structurally, the 26S proteasome is a 2.5 MDa molecular machine comprising of a barrel-like 20S proteolytic core particle (CP) and one or two 19S regulatory particles (RP/PA700) at the either end. Functionally, it plays a key role in the cell's decision to proliferate or undergo

R. Verma
Department of Biology, Howard Hughes Medical Institute, California Institute of Technology, Pasadena, CA 91125, USA
e-mail: verma@caltech.edu

K. Sakamoto and E. Rubin (eds.), *Modulation of Protein Stability in Cancer Therapy*,
DOI: 10.1007/978-0-387-69147-3_2, © Springer Science+Business Media, LLC 2009

apoptosis by degrading proteins such as the cyclins, cyclin-dependent kinase inhibitors, p53, IκBα, Bax, Bcl-2, to name just a few (reviewed in Richardson et al. 2005a). The repertoire of substrates is only expected to grow as we learn more about the pathways that feed into the Ub-proteasome system (UPS). Interestingly, some of the above regulatory substrates that undergo signal-induced, Ub-dependent turnover can also be degraded by the 20S proteasome independently of Ub, and in some cases, even ATP (reviewed in Verma and Deshaies 2000; Zetter and Mangold 2005). As discussed below in this chapter, this has therapeutic implications.

The ATP- and Ub-Dependent Degradation Reaction

Proteolysis of a typical labile substrate of the 26S proteasome is believed to involve the following steps: (a) the substrate is "marked" with a polyUb chain through the concerted action of a "Ub bucket brigade" involving an enzymatic cascade of E1 (Ub-activating enzyme), E2 (Ub-conjugating enzyme) and E3 (Ub-ligase) activities (reviewed in Hershko et al. 2000). (b) Following ubiquitination, the substrate is recognized by either the intrinsic receptor of the 26S proteasome such as Rpn10/S5a, or "shuttle" receptors such as Rad23/hHR23A/B, Dsk2/Ubiquilin, or adaptors of the AAA-ATPase Cdc48/p97 (reviewed in Elsasser and Finley 2005; Verma et al. 2004a). (c) Once bound to the proteasome, the substrate is unfolded by the resident hexameric ATPase (Rpt) ring while simultaneously being deubiquitinated by the JAMM-domain containing metalloprotease Rpn11/Poh1/S13 (Verma et al. 2002; Yao and Cohen 2002). Unfolding with concomitant deubiquitination is an essential step in the degradation reaction because the opening of the 20S CP orifice is too narrow for the entry of globular proteins with attached Ub chains. (d) As the substrate is threaded into the CP, it is degraded into peptides by the sequestered peptidase activities of the subunits, while Ub is recycled (Kisselev et al. 2006).

Targeting the 20S Core Proteases: The Approval of Bortezomib for Multiple Myeloma and Validation of the Proteasomal Proteases as a Therapeutic Target

The 20S CP consists of four seven-membered rings stacked on each other to generate a sequestered chamber for the resident peptidase activities. The outer two rings containing αsubunits are catalytically inactive. Moreover, the tails of the three of these α subunits form a "gate" that blocks substrate access (Groll et al. 2000). The 20S CP therefore exists in an auto-inhibited state which has to be activated before unfolded substrates can enter the inner chamber. The C-terminal peptides of two of the 19S RP ATPases (Rpt2 and Rpt5) can induce the opening of the gate (Smith et al. 2007). Once in, the unfolded protein is exposed to the proteolytic milieu of the inner catalytic chamber composed of two rings of seven β subunits. Peptides are generated by the concerted action of the chymotrypsin-like, trypsin-like, and caspase-like activities of the β-5, β-2, β-1 subunits respectively. Mammalian CP can additionally cleave after the branched chain and small neutral amino acids (Kisselev et al. 2006).

The 20S CP protease is different from all cellular proteases in that each proteasome active site uses the side chain hydroxyl group of an amino-terminal threonine as the catalytic nucleophile. Accordingly, inhibitors that target this moiety have been identified. The first generation of peptide aldehyde (Vinitsky et al. 1992), vinyl sulfone (Bogyo et al. 1997) inhibitors were followed by the more specific peptide boronic acid inhibitors (Adams et al. 1998). The best characterized amongst the latter group is bortezomib (VELCADE, or PS-341, Millennium Pharmaceuticals, Inc) which reversibly inhibits the chymotryptic activity of the proteasome and leads to the death of the cell (Adams 2004). Bortezomib has proved effective for multiple myeloma (MM) (Richardson et al. 2005b) and some forms of non-Hodgkin's lymphoma (NHL) (Goy et al. 2005; O'Connor et al. 2005). It was approved by the FDA in 2003 for the treatment of MM, and in 2006 for the treatment of mantle cell lymphoma.

As with all drugs, some patients have proved refractory to bortezomib treatment, whereas others have discontinued treatment due to side effects, prompting the development of a third generation of proteasome inhibitors (reviewed in Voorhees and Orlowski 2006). Two irreversible proteasome inhibitors are currently under development: (a) Salinosporamide A (NPI-0052) from marine actinocyte bacteria that induces apoptosis of MM and CLL tumor cells (Chauhan et al. 2005; Macherla et al. 2005). (b) PR-171, an epoxy-ketone peptidyl inhibitor (Demo et al. 2007) related to the natural product epoxomicin which was initially identified as an anti-tumor agent (Hanada et al. 1992), was later shown to be a specific inhibitor of the proteasome (Kim et al. 1999). Besides epoxomicin and salinosporamide, there are other natural compounds such as (−)-epigallocatechin 3-gallate ((−)-EGCG), the most abundant catechin in green tea that acts as an anticancer agent and can inhibit the chymotrypsin-like activity of purified 20S CP (Smith et al. 2002). Analogs of (−)-EGCG, such as a benzilate derivative have been patented. For a summary of all patented small molecule inhibitors of the UPS, see (Guedat and Colland 2007).

Alternative Therapeutic Targets within the 26S Proteasome

Proteasomes are ubiquitous and abundant entities, comprising of the primary non-lysosomal proteolytic pathway of the cell. Therefore, it was predicted that proteasomal inhibitors would affect both normal, as well as transformed cells. However, preclinical studies have indeed demonstrated increased susceptibility of tumor cells to proteasome-inhibitor mediated apoptosis (reviewed in Adams 2004). The molecular basis for this selective killing is not yet fully understood because both pro- and anti-apoptotic proteins are proteasomal substrates. However, the proliferative status of the cell, net inhibition of NFκB activity, and the microenvironment of the tumor cell itself must all impact in determining the final therapeutic index of the inhibitor. The last-mentioned property may also explain why hematologic tumors have been more successfully treated thus far. Thus continued research into other possible sites of proteasomal intervention would be deemed prudent (Fig. 1), and these are described below.

(a) Targeting the Receptor Pathway.
(b) Targeting the JAMM domain-containing deubiquitinating enzyme (DUB) RPN11/POH1/S13 and other proteasomal DUBs.
(c) Targeting the ATPases.
(d) Targeting the ATPase-20S core interface.

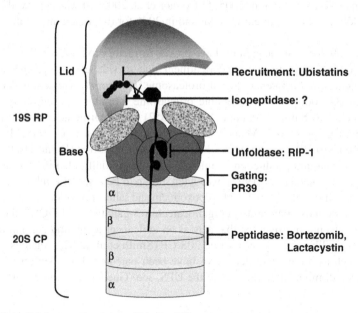

Fig. 1 Potential therapeutic targets within the 26S proteasome.
The 26S proteasome is comprised of a 20S core particle (20S CP) that is capped of at the ends by the 19S regulatory particle (19S RP). The peptidase activity is sequestered within the 20S CP. Traditional proteasome inhibitors such as Bortezomid inhibit this step. The 19S RP can be subdivided into the Lid and Base. Ubiquitinated substrates of the proteasome (black), are recognized by receptor proteins that are localized at either the hinge of the Lid and the Base, or bound to scaffold proteins in the Base (stippled grey). This recruitment step is inhibited by Ubistatins. Once bound to the proteasome, the substrate is unfolded by the resident ATPase hexamer, and threaded into the 20S CP. The ubiquitin chain (black) is concomitantly cleaved by an isopeptidase in the Lid. RIP-1 is an inhibitor postulated to inhibit unfolding by the ATPase, while PR39 could block the opening of the 20S CP cavity

Targeting the Receptor Pathway

(a) *Ubistatins*. Polyubiquitinated proteins are bound by UBA-, UIM-, and UBX-containing proteins and targeted to the 26S proteasome for degradation. Since this is a non-enzymatic step, it was believed to be a non-druggable step. However, a small molecule screen for inhibitors of proteolysis of cyclin and β-catenin reporters using staged Xenopus extracts resulted in the identification of a class of compounds designated Ubistatins that inhibited proteolysis by disrupting binding between the receptor and the polyUb chain. A new class of inhibitors was thus identified using this chemical genetic screen (Verma et al. 2004b). Ubistatins are more potent than MG132 and lactacystins in Xenopus extracts, as well as yeast extracts (Verma et al., unpublished data), presumably because the latter inhibitors are rapidly hydrolyzed in the extracts. However, their use is limited to in vitro analyses because their charge renders the cell membrane impermeable.

(b) The UIM-containing protein S5a is the intrinsic proteasomal receptor in mammalian cells. It has recently been reported that ectopic expression of either S5aC (a C-terminal truncation), or S5a-UIM resulted in A549 lung cancer cell death, whereas non-cancerous BEAS-2B cells were unaffected. As reported for ubistatins, accumulation of polyubiquitinated proteins was observed (Elangovan et al. 2007). Another inhibitor that most likely operates by the same pathway mechanistically is an artificial tandem Ub construct (tUb6), designed to be insensitive to intracellular Ub hydrolases. When over-expressed in vivo, this construction inhibits growth of budding yeast, while injection of the tUb6 mRNA into Xenopus embryos results in inhibition of the mitotic cell division (Saeki et al. 2005).

Targeting the JAMM Domain of the Proteasomal DUB Rpn11/Poh1/S13 and Other Proteasomal Isopeptidases

Rpn11 contains a JAMM/MPN + motif sequence $Ex_nHxHx_7Sx_2D$ that resembles the active site of zinc metalloproteases. Resolution of the crystal structure of an archaebacterium JAMM-containing protein confirmed that zinc is coordinated by the two histidines and aspartic acid of the JAMM motif (Ambroggio et al. 2004; Tran et al. 2003). Rpn11 catalyzes the hydrolysis of the proximal Ub isopeptide bond to the substrate (Verma et al. 2002; Yao and Cohen 2002). Mutation of the active site histidines is lethal both in yeast and fly (Lundgren et al. 2003), and it has subsequently been shown that the JAMM motif of human/Poh1/S13 is also essential for HeLa cell viability (Gallery et al. 2007).

Besides Rpn11/Poh1, two other nonessential DUBs (Uch37 and Usp14) are associated with the mammalian 26S proteasome (Koulich et al. 2007). They belong to the cysteine protease class and are thus distinct from metalloisopeptidases (reviewed in Love et al. 2007). Uch37 and Usp14 are believed to help process the Ub chain (amongst other postulated functions), resulting in the generation of free Ub monomers. Although no specific inhibitors have been identified for any of these DUBs per se, general small molecule inhibitors of isopeptidases have been identified using Ub-PEST and z-LRGG-AMC as substrates. These inhibitors cause cell death of colon cancer cells with the accumulation of ubiquitinated proteins (Mullally et al. 2001).

Targeting the ATPases

The base of the 19S RP is comprised of six ATPases (Rpt1-6) that dock directly atop the 20S CP. The ATPases are not equivalent in function (Rubin et al. 1998), but collectively, the following three functions have been ascribed to the hexameric ring (a) they can bind tetraUb chains (b) they unfold substrates (c) they are involved in "gating" the 20S CP (reviewed in Saeki and Tanaka 2007). A recently described inhibitor: Regulatory Particle Inhibitor Peptoid 1 (RIP-1) has been shown to target Rpt4/Sug2. It inhibited the unfolding activity of 19S RP in vitro, and inhibited p27 turnover in Hela cells in vivo (Lim et al. 2007).

Targeting the ATPase-20S Core Interface

PR39 is a natural proline- and arginine-rich 39 amino acid antibacterial peptide derived from porcine bone marrow that selectively inhibits the turnover of HIF-1α (Li et al. 2000) and IκBα (Gao et al. 2000). Mechanistically, PR39 is postulated to interact with the non-catalytic α7 subunit of the 20S CP, with consequent changes in proteasome conformation such that switching between the open and closed conformations is prevented, with consequent inhibition of the turnover of selective substrates (Gaczynska et al. 2003). In vivo, inflammatory responses are inhibited, but angiogenesis is stimulated.

Ub-Independent Turnover by 26S Proteasomes

The most well studied example in this class of proteins is ornithine decarboxylase (ODC), the activity of which is elevated in most forms of human tumors. ODC activity is regulated by the protein antizyme 1 (AZ1), which functions as a tumor-suppressor by binding and inactivating ODC. In addition, AZ1 promotes the degradation of ODC by enhancing its binding to the 26S proteasome (Hoyt and Coffino 2004). ODC binds the proteasome via a C-terminal 37 amino acid native degron which is transposable (Takeuchi et al. 2007). Amazingly, this native degron shares molecular determinants with a polyUb chain, because AZ1 stimulated degradation of ODC can be competed by a polyUb chain (Zhang et al. 2003)! Interestingly, ubistatin A did not inhibit turnover of ODC, whereas ubistatin B did indeed block turnover of ODC by purified 26S proteasomes in vitro (Verma et al. 2004b).

Ub- and ATP-Independent Turnover by 20S Proteasomes

A number of regulatory proteins that are known to undergo signal-induced, Ub-dependent turnover, can bypass E3 Ub-Ligases and be degraded without any energy expenditure. This bypass mechanism may represent constitutive turnover of the target protein, or turnover of the target lacking its binding partners. Whatever the underlying physiological cue, it has recently been demonstrated that proteins such as p21 (Waf/Cip1) and steroid receptor coactivator SRC-3, can be degraded in the absence of ATP, and a Ub targeting signal by the REGγ-20S complex (Chen et al. 2007; Li et al. 2007).

As discussed above, 20S proteasomes typically exist in an autoinhibited state and are activated by the docking of the 19S RP. They can additionally be activated by another complex: the 11S PA28/REG complex. There are three PA28/REG homologs called α, β, and γ. They can exist as cytoplasmic heteroheptameric PA28α/β complexes, or as nuclear homoheptamers of PA28/REGγ. The degradation of proteins involved in proliferation by REGγ-20S provide a molecular basis to the earlier findings that PA28γ-deficient mice grow slowly and are smaller than wild-type mice at maturity. PA28γ$^{-/-}$ fibroblasts exhibit increased levels of apoptosis, and high levels of PA28γ are found in thyroid cancers. This form of the proteasome is also enriched in the brain (reviewed in Rechsteiner and Hill 2005).

Besides the REGγ-20S complex, it has been reported that the 20S CP alone can degrade p53 and p73 in an ATP- and Ub-independent reaction (Asher et al. 2002). The cytosolic quinone oxidoreductases NQO1 and NQO2 interact with p53, and protect it from degradation by 20S CP (Gong et al. 2007). Although null mice of the respective oxidoreductases reach maturity, they develop myelogenous hyperplasia of bone marrow. In vitro inhibition of NQO1 activity by dicoumarol induces p53 proteasomal degradation. Humans carrying a polymorphic inactive NQO1 are at increased risk to develop adult leukemia (Smith et al. 2001).

Proteolytic and Non-Proteolytic Roles of the Proteasome in Chromatin Biology

DNA-dependent processes such as replication and transcription involve multi-protein complexes, individual members of which may be proteolytic targets of the 26S proteasome (reviewed in Lipford and Deshaies 2003). Indeed, degradation of some activators as promoters has been correlated with activation of gene expression, with degradation facilitating multiple rounds of transcription initiation (Dennis et al. 2005; Lipford et al. 2005). The 26S proteasome also plays a surveillance role by preventing incorrect transcription on specific regulatory regions in embryonic stem cells (Szutorisz et al. 2006). Evidence also exists for non-proteolytic roles of the proteasome in transcription and DNA repair (reviewed in Collins and Tansey 2006). Typically, these functions involve proteasomal subcomplexes. For example, the 19S RP has been shown to promote recruitment of the SAGA complex onto chromatin, and enhance its histone acetyltransferase activity (Lee et al. 2005). Since the SAGA complex has also been implicated in transcription elongation, mRNA export, and nucleotide excision repair, the 19S subcomplex could potentially regulate all of these processes, primarily through its polyUb-binding and chaperone activities (reviewed in Baker and Grant 2007).

The proteasome, as outlined above, plays a role in multiple cellular pathways. This property would suggest that inhibiting its proteolytic activity would have been detrimental to both malignant, and non-malignant cells. However, proteasomal protease inhibitors do indeed kill tumor cells selectively, and significant efficacy has been demonstrated against MM and NHL, both hematological tumors. New avenues of research that have opened up based on these pioneering findings include the testing of these inhibitors, either singly, or in combination, against solid tumors of different tissue types. Given the myriad functions of the proteasome, toxicity will remain an issue, so a consideration of other points of intervention within the 26S proteasome would be prudent. For example, inhibiting the initial recruitment step might prove less toxic than inhibiting proteolytic activity because the former would target only Ub-dependent pathways whereas the latter blocks all pathways. Similarly, given the prevalence of REGγ in brain tissue, it may prove fruitful to identify inhibitors of the REGγ-CP pathway and test them against neurological tumors. It is hoped that such studies will enhance our basic understanding of the proteasome, and at the same time lead to the development of anti-cancer drugs that prolong life without significantly affecting its quality.

References

Adams, J. (2004). The proteasome: a suitable antineoplastic target. Nat Rev Cancer 4, 349–360.

Adams, J., Behnke, M., Chen, S., Cruickshank, A.A., Dick, L.R., Grenier, L., Klunder, J.M., Ma, Y.T., Plamondon, L., and Stein, R.L. (1998). Potent and selective inhibitors of the proteasome: dipeptidyl boronic acids. Bioorg Med Chem Lett 8, 333–338.

Ambroggio, X.I., Rees, D.C., and Deshaies, R.J. (2004). JAMM: a metalloprotease-like zinc site in the proteasome and signalosome. PLoS Biol 2, E2.

Asher, G., Lotem, J., Sachs, L., Kahana, C., and Shaul, Y. (2002). Mdm-2 and ubiquitin-independent p53 proteasomal degradation regulated by NQO1. Proc Natl Acad Sci U S A 99, 13125–13130.

Baker, S.P., and Grant, P.A. (2007). The SAGA continues: expanding the cellular role of a transcriptional co-activator complex. Oncogene 26, 5329–5340.

Bogyo, M., McMaster, J.S., Gaczynska, M., Tortorella, D., Goldberg, A.L., and Ploegh, H. (1997). Covalent modification of the active site threonine of proteasomal beta subunits and the Escherichia coli homolog HslV by a new class of inhibitors. Proc Natl Acad Sci U S A 94, 6629–6634.

Chauhan, D., Catley, L., Li, G., Podar, K., Hideshima, T., Velankar, M., Mitsiades, C., Mitsiades, N., Yasui, H., Letai, A., et al. (2005). A novel orally active proteasome inhibitor induces apoptosis in multiple myeloma cells with mechanisms distinct from Bortezomib. Cancer Cell 8, 407–419.

Chen, X., Barton, L.F., Chi, Y., Clurman, B.E., and Roberts, J.M. (2007). Ubiquitin-independent degradation of cell-cycle inhibitors by the REGgamma proteasome. Mol Cell 26, 843–852.

Collins, G.A., and Tansey, W.P. (2006). The proteasome: a utility tool for transcription? Curr Opin Genet Dev 16, 197–202.

Demo, S.D., Kirk, C.J., Aujay, M.A., Buchholz, T.J., Dajee, M., Ho, M.N., Jiang, J., Laidig, G.J., Lewis, E.R., Parlati, F., et al. (2007). Antitumor activity of PR-171, a novel irreversible inhibitor of the proteasome. Cancer Res 67, 6383–6391.

Dennis, A.P., Lonard, D.M., Nawaz, Z., and O'Malley, B.W. (2005). Inhibition of the 26S proteasome blocks progesterone receptor-dependent transcription through failed recruitment of RNA polymerase II. J Steroid Biochem Mol Biol 94, 337–346.

Elangovan, M., Choi, E.S., Jang, B.G., Kim, M.S., and Yoo, Y.J. (2007). The ubiquitin-interacting motif of 26S proteasome subunit S5a induces A549 lung cancer cell death. Biochem Biophys Res Commun 364, 226–230.

Elsasser, S., and Finley, D. (2005). Delivery of ubiquitinated substrates to protein-unfolding machines. Nat Cell Biol 7, 742–749.

Gaczynska, M., Osmulski, P.A., Gao, Y., Post, M.J., and Simons, M. (2003). Proline- and arginine-rich peptides constitute a novel class of allosteric inhibitors of proteasome activity. Biochemistry 42, 8663–8670.

Gallery, M., Blank, J.L., Lin, Y., Gutierrez, J.A., Pulido, J.C., Rappoli, D., Badola, S., Rolfe, M., and Macbeth, K.J. (2007). The JAMM motif of human deubiquitinase Poh1 is essential for cell viability. Mol Cancer Ther 6, 262–268.

Gao, Y., Lecker, S., Post, M.J., Hietaranta, A.J., Li, J., Volk, R., Li, M., Sato, K., Saluja, A.K., Steer, M.L., et al. (2000). Inhibition of ubiquitin-proteasome pathway-mediated I kappa B alpha degradation by a naturally occurring antibacterial peptide. J Clin Invest 106, 439–448.

Gong, X., Kole, L., Iskander, K., and Jaiswal, A.K. (2007). NRH:quinone oxidoreductase 2 and NAD(P)H:quinone oxidoreductase 1 protect tumor suppressor p53 against 20s proteasomal degradation leading to stabilization and activation of p53. Cancer Res 67, 5380–5388.

Goy, A., Younes, A., McLaughlin, P., Pro, B., Romaguera, J.E., Hagemeister, F., Fayad, L., Dang, N.H., Samaniego, F., Wang, M., et al. (2005). Phase II study of proteasome inhibitor bortezomib in relapsed or refractory B-cell non-Hodgkin's lymphoma. J Clin Oncol 23, 667–675.

Groll, M., Bajorek, M., Kohler, A., Moroder, L., Rubin, D.M., Huber, R., Glickman, M.H., and Finley, D. (2000). A gated channel into the proteasome core particle. Nat Struct Biol 7, 1062–1067.

Guedat, P., and Colland, F. (2007). Patented small molecule inhibitors in the ubiquitin proteasome system. BMC Biochem 8(Suppl 1), S14.

Hanada, M., Sugawara, K., Kaneta, K., Toda, S., Nishiyama, Y., Tomita, K., Yamamoto, H., Konishi, M., and Oki, T. (1992). Epoxomicin, a new antitumor agent of microbial origin. J Antibiot (Tokyo) 45, 1746–1752.

Hershko, A., Ciechanover, A., and Varshavsky, A. (2000). Basic Medical Research Award. The ubiquitin system. Nat Med 6, 1073–1081.

Hoyt, M.A., and Coffino, P. (2004). Ubiquitin-free routes into the proteasome. Cell Mol Life Sci 61, 1596–1600.

Kim, K.B., Myung, J., Sin, N., and Crews, C.M. (1999). Proteasome inhibition by the natural products epoxomicin and dihydroeponemycin: insights into specificity and potency. Bioorg Med Chem Lett 9, 3335–3340.

Kisselev, A.F., Callard, A., and Goldberg, A.L. (2006). Importance of the different proteolytic sites of the proteasome and the efficacy of inhibitors varies with the protein substrate. J Biol Chem 281, 8582–8590.

Koulich, E., Li, X., and Demartino, G.N. (2007). Relative structural and functional roles of multiple deubiquitylating proteins associated with mammalian 26S proteasome. Mol Biol Cell 19(3):1072–1082.

Lee, D., Ezhkova, E., Li, B., Pattenden, S.G., Tansey, W.P., and Workman, J.L. (2005). The proteasome regulatory particle alters the SAGA coactivator to enhance its interactions with transcriptional activators. Cell 123, 423–436.

Li, J., Post, M., Volk, R., Gao, Y., Li, M., Metais, C., Sato, K., Tsai, J., Aird, W., Rosenberg, R.D., et al. (2000). PR39, a peptide regulator of angiogenesis. Nat Med 6, 49–55.

Li, X., Amazit, L., Long, W., Lonard, D.M., Monaco, J.J., and O'Malley, B.W. (2007). Ubiquitin- and ATP-independent proteolytic turnover of p21 by the REGgamma-proteasome pathway. Mol Cell 26, 831–842.

Lim, H.S., Cai, D., Archer, C.T., and Kodadek, T. (2007). Periodate-triggered cross-linking reveals Sug2/Rpt4 as the molecular target of a peptoid inhibitor of the 19S proteasome regulatory particle. J Am Chem Soc 129, 12936–12937.

Lipford, J.R., and Deshaies, R.J. (2003). Diverse roles for ubiquitin-dependent proteolysis in transcriptional activation. Nat Cell Biol 5, 845–850.

Lipford, J.R., Smith, G.T., Chi, Y., and Deshaies, R.J. (2005). A putative stimulatory role for activator turnover in gene expression. Nature 438, 113–116.

Love, K.R., Catic, A., Schlieker, C., and Ploegh, H.L. (2007). Mechanisms, biology and inhibitors of deubiquitinating enzymes. Nat Chem Biol 3, 697–705.

Lundgren, J., Masson, P., Realini, C.A., and Young, P. (2003). Use of RNA interference and complementation to study the function of the *Drosophila* and human 26S proteasome subunit S13. Mol Cell Biol 23, 5320–5330.

Macherla, V.R., Mitchell, S.S., Manam, R.R., Reed, K.A., Chao, T.H., Nicholson, B., Deyanat-Yazdi, G., Mai, B., Jensen, P.R., Fenical, W.F., et al. (2005). Structure-activity relationship studies of salinosporamide A (NPI-0052), a novel marine derived proteasome inhibitor. J Med Chem 48, 3684–3687.

Mullally, J.E., Moos, P.J., Edes, K., and Fitzpatrick, F.A. (2001). Cyclopentenone prostaglandins of the J series inhibit the ubiquitin isopeptidase activity of the proteasome pathway. J Biol Chem 276, 30366–30373.

O'Connor, O.A., Wright, J., Moskowitz, C., Muzzy, J., MacGregor-Cortelli, B., Stubblefield, M., Straus, D., Portlock, C., Hamlin, P., Choi, E., et al. (2005). Phase II clinical experience with the novel proteasome inhibitor bortezomib in patients with indolent non-Hodgkin's lymphoma and mantle cell lymphoma. J Clin Oncol 23, 676–684.

Pickart, C.M., and Cohen, R.E. (2004). Proteasomes and their kin: proteases in the machine age. Nat Rev Mol Cell Biol 5, 177–187.

Rechsteiner, M., and Hill, C.P. (2005). Mobilizing the proteolytic machine: cell biological roles of proteasome activators and inhibitors. Trends Cell Biol 15, 27–33.

Richardson, P.G., Mitsiades, C., Hideshima, T., and Anderson, K.C. (2005a). Proteasome inhibition in the treatment of cancer. Cell Cycle 4, 290–296.

Richardson, P.G., Sonneveld, P., Schuster, M.W., Irwin, D., Stadtmauer, E.A., Facon, T., Harousseau, J.L., Ben-Yehuda, D., Lonial, S., Goldschmidt, H., et al. (2005b). Bortezomib or high-dose dexamethasone for relapsed multiple myeloma. N Engl J Med 352, 2487–2498.

Rubin, D.M., Glickman, M.H., Larsen, C.N., Dhruvakumar, S., and Finley, D. (1998). Active site mutants in the six regulatory particle ATPases reveal multiple roles for ATP in the proteasome. EMBO J 17, 4909–4919.

Saeki, Y., and Tanaka, K. (2007). Unlocking the proteasome door. Mol Cell 27, 865–867.

Saeki, Y., Isono, E., Shimada, M., Kawahara, H., Yokosawa, H., and Toh, E.A. (2005). Knocking out ubiquitin proteasome system function in vivo and in vitro with genetically encodable tandem ubiquitin. Methods Enzymol 399, 64–74.

Smith, D.M., Wang, Z., Kazi, A., Li, L.H., Chan, T.H., and Dou, Q.P. (2002). Synthetic analogs of green tea polyphenols as proteasome inhibitors. Mol Med 8, 382–392.

Smith, D.M., Chang, S.C., Park, S., Finley, D., Cheng, Y., and Goldberg, A.L. (2007). Docking of the proteasomal ATPases' carboxyl termini in the 20S proteasome's alpha ring opens the gate for substrate entry. Mol Cell 27, 731–744.

Smith, M.T., Wang, Y., Kane, E., Rollinson, S., Wiemels, J.L., Roman, E., Roddam, P., Cartwright, R., and Morgan, G. (2001). Low NAD(P)H:quinone oxidoreductase 1 activity is associated with increased risk of acute leukemia in adults. Blood 97, 1422–1426.

Szutorisz, H., Georgiou, A., Tora, L., and Dillon, N. (2006). The proteasome restricts permissive transcription at tissue-specific gene loci in embryonic stem cells. Cell 127, 1375–1388.

Takeuchi, J., Chen, H., and Coffino, P. (2007). Proteasome substrate degradation requires association plus extended peptide. EMBO J 26, 123–131.

Tran, H.J., Allen, M.D., Lowe, J., and Bycroft, M. (2003). Structure of the Jab1/MPN domain and its implications for proteasome function. Biochemistry 42, 11460–11465.

Verma, R., and Deshaies, R.J. (2000). A proteasome howdunit: the case of the missing signal. Cell 101, 341–344.

Verma, R., Aravind, L., Oania, R., McDonald, W.H., Yates, J.R., III, Koonin, E.V., and Deshaies, R.J. (2002). Role of Rpn11 metalloprotease in deubiquitination and degradation by the 26S proteasome. Science 298, 611–615.

Verma, R., Oania, R., Graumann, J., and Deshaies, R.J. (2004a). Multiubiquitin chain receptors define a layer of substrate selectivity in the ubiquitin-proteasome system. Cell 118, 99–110.

Verma, R., Peters, N.R., D'Onofrio, M., Tochtrop, G.P., Sakamoto, K.M., Varadan, R., Zhang, M., Coffino, P., Fushman, D., Deshaies, R.J., et al. (2004b). Ubistatins inhibit proteasome-dependent degradation by binding the ubiquitin chain. Science 306, 117–120.

Vinitsky, A., Michaud, C., Powers, J.C., and Orlowski, M. (1992). Inhibition of the chymotrypsin-like activity of the pituitary multicatalytic proteinase complex. Biochemistry 31, 9421–9428.

Voorhees, P.M., and Orlowski, R.Z. (2006). The proteasome and proteasome inhibitors in cancer therapy. Annu Rev Pharmacol Toxicol 46, 189–213.

Yao, T., and Cohen, R.E. (2002). A cryptic protease couples deubiquitination and degradation by the proteasome. Nature 419, 403–407.

Zetter, B.R., and Mangold, U. (2005). Ubiquitin-independent degradation and its implication in cancer. Future Oncol 1, 567–570.

Zhang, M., Pickart, C.M., and Coffino, P. (2003). Determinants of proteasome recognition of ornithine decarboxylase, a ubiquitin-independent substrate. EMBO J 22, 1488–1496.

Targeting HAUSP: Killing Two Birds with One Stone

Christopher L. Brooks and Wei Gu

Abstract The cysteine protease deubiquitinase herpesvirus-associated ubiquitin-specific protease (HAUSP) has been fairly well characterized in recent years and has emerged as an important component of the p53-Mdm2-MdmX signaling pathway. HAUSP was first identified as a 135 kD cellular factor associated with the herpesvirus regulatory protein Vmw110. Interestingly, reduction or ablation of HAUSP leads to the DNA-damage-induced MdmX degradation and instability of Mdm2, leading to a robust stabilization of p53. Removal of HAUSP reduces the half-life of Mdm2 from 30 min to approximately 5 min. These four proteins, p53, Mdm2, MdmX, and HAUSP have a delicate, intimate balance that maintains p53 proteins levels and is critical for an appropriate DNA damage response to occur. And because of the strong p53 induction that occurs in the absence of HAUSP, it represents a potential chemotherapeutic target for the stabilization and activation of p53 in cells that retain a wildtype copy of the gene.

Keywords p53 • Mdm2 • HAUSP • Destabilization • Ubiquitination • Deubiquitination • Activation and stability

Introduction

Deubiquitination enzymes (DUBs) have been shown to function in a wide variety of cellular processes (Hershko and Ciechanover 1998; Laney and Hochstrasser 1999; Pickart and Eddins 2004). Still the number of characterized DUBs to date remains low when compared to the total number of 95 putative DUBs that exist within the larger class of cellular proteases in the human genome (Puente and

C.L. Brooks and W. Gu
Department of Pathology, Institute for Cancer Genetics, College of Physicians & Surgeons, Columbia University, 1130 St. Nicholas Ave, New York, NY, 10032, USA
e-mail: wg8@columbia.edu

K. Sakamoto and E. Rubin (eds.), *Modulation of Protein Stability in Cancer Therapy*,
DOI: 10.1007/978-0-387-69147-3_3, © Springer Science+Business Media, LLC 2009

Lopez-Otin 2004). DUBs can be sub-divided into five distinct classes including ubiquitin specific proteases (USPs), ubiquitin C-terminal hydrolases (UCHs), ovarian tumor proteases (OTUs), JAMM Motif proteases, and Machado Joseph disease protein domain proteases (MJDs) (Amerik and Hochstrasser 2004; Millard and Wood 2006). The largest class of DUBs is the ubiquitin specific proteases, a class of 50 proteins that are involved in diverse cellular processes and antagonize the ubiquitin-proteosome pathway (Fig. 1a) (Nijman et al. 2005). The process of removing ubiquitin moieties is a critical regulatory step for cellular proteins that require precise spatial and temporal control at all times. Without this level of control, improperly ubiquitinated proteins and/or reduction of protein levels may lead to deleterious signaling consequences for the cell.

An overview of all DUBs has been well reviewed in several recent papers (Hershko and Ciechanover 1998; Nijman et al. 2005). Here we wish to discuss recent advances and implications made for one particular cysteine protease DUB, the herpesvirus-associated ubiquitin-specific protease (HAUSP). HAUSP has been shown to deubiquitinate several substrates, three of them being critical factors in the p53 signaling pathway (Brooks et al. 2007). This pathway has been the focus of intense research over the past 18 years as it is frequently mutated in a large percentage of human tumors. However, of the tumors that retain wildtype p53, an emphasis has been placed on finding compounds that block its direct negative regulator, the E3 ligase Mdm2. Since HAUSP exhibits enzymatic activity towards both p53 and Mdm2, we suggest that HAUSP may be an equally logical target for activating p53 in this subset of tumors.

HAUSP: Discovery, Structure, and Targets

HAUSP was originally cloned and identified in 1997 as a 135 kD cellular protein that associated with the Herpesvirus regulatory protein Vmw110, a subset of which co-localized with PML nuclear bodies (Everett et al. 1997). The interaction between HAUSP and Vmw110 was shown to be important for the activation of herpes simplex virus type 1 expression and replication (Everett et al. 1999). Located on chromosome 16p13.3, HAUSP is a cysteine protease DUB consisting of 1,102 amino acids (Robinson et al. 1998). It can be loosely divided into three structural and functional domains: an N-terminal TRAF-like domain, a Cys-His catalytic domain, and a C-terminal domain (Fig. 1b). Other specific functional domains include a conserved cysteine-histidine catalytic domain between amino acids 203 and 608, a meprin and TRAF homology domain (MATH) between amino acids 53 and 208, and several protein binding domains. The crystal structure of the catalytic domain shows a three-domain structure consisting of a fingers and palm-thumb scaffold (Brooks et al. 2007). When the structure associates with ubiquitin aldehyde, a conformational change occurs that realigns the catalytic triad for proper positioning of substrate catalysis. This change in conformation when bound to ubiquitin is a common feature among UCHs and USPs and prevents promiscuous

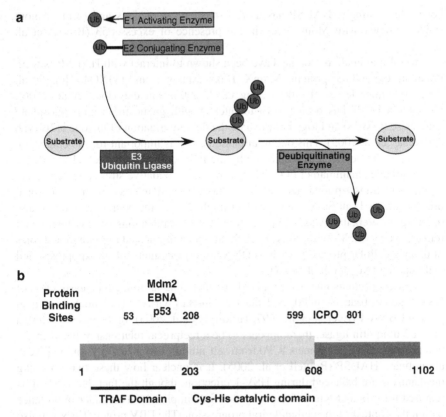

Fig. 1 Schematic representation of HAUSP and its functions. (**a**) An ubiquitin molecule is first activated by an E1 activating enzyme through the ATP-dependent formation of a covalent intermediate with ubiquitin at its active site. The activated ubiquitin is then transferred to the active site of an E2 conjugating molecule. Substrate specificity is provided for by the E3 ubiquitin ligase, which transfers the activated ubiquitin from the E2 to the target lysine on the substrate. (**b**) A schematic representation of the HAUSP protein structure. HAUSP consists of 1,102 amino acids and can be divided into three specific domains: N-terminus TRAF-like domain, the cysteine-histidine catalytic domain, and the C-terminus domain

protease activity against random substrates. However, the exact mechanisms for regulating HAUSP functions remain unclear. It has recently been shown that HAUSP can be both ubiquitinated and neddylated, though the consequences of these posttranslational modifications remain unclear (Lee et al. 2005). When N-terminal cocrystal structures were solved for HAUSP together with Mdm2, p53, and EBNA peptides, HAUSP was shown to have a binding preference for the consensus sequence motif P/AXXS (Sheng et al. 2006). These three proteins all compete for a binding region within a pocket of N-terminus MATH domain. The mutually exclusive manner of binding, particularly between HAUSP-p53 and HAUSP-Mdm2, has important implications in p53 pathway regulation, as discussed below. Interestingly, HAUSP has more extensive interactions and forms a more stable complex with Mdm2 even in the presence of excess p53 (Brooks et al. 2007).

Several mammalian proteins have been shown to interact with HAUSP as well, including the cellular proteins MdmX, Daxx, Ataxin-1, and FOXO4 (Hong et al. 2002; Meulmeester et al. 2005; Tang et al. 2006; van der Horst et al. 2006). Drosophila USP7 has been shown to interact with guanosine 5-monophosphate synthetase (GMPS) and together mediate the deubiquitination of histone H2B (van der Knaap et al. 2005). This deubiquitination event is important for the epigenetic silencing of homeotic genes mediated by the Polycomb protein (Pc). HAUSP also deubiquitinates mammalian FOXO4 in response to oxidative stress and negatively regulates its transcriptional activity. In addition, HAUSP interacts with the neuronal protein Ataxin-1, an SCA1 gene product implicated in the neurodegenerative disorder spinocerebellar Ataxia Type-1, though the implications of this interaction remain to be seen. More recently, HAUSP has been implicated in more global roles of diverse cellular processes such as DNA replication, metabolism, apoptosis, and vial budding (Kessler et al. 2007).

Two viral proteins interact with HAUSP, including an immediate early class of HSV-1 gene products, ICP0, and the Epstein-Barr virus (EBV) nuclear antigen protein EBNA-1 (Everett et al. 1997; Holowaty et al. 2003). In the case of ICP0, a viral E3 ubiquitin ligase, there appears to be a reciprocal relationship between the two proteins. HAUSP protects ICP0 from self-ubiquitination activity, and ICP0 can ubiquitinate HAUSP (Boutell et al. 2005). It is unclear how these two opposing mechanisms are balanced during HSV-1 infection, though the fact that HAUSP is required for efficient lytic infection by the virus implicates its broader importance within the context of the cell and viral progression. The EBV protein EBNA-1 also has a high binding affinity for HAUSP, and binding between these two proteins disrupts the HAUSP-p53 interaction (Sheng et al. 2006). Together these recent data suggest that viruses may exploit a cellular factor that has a close, intimate relationship with p53. By binding to HAUSP, or targeting it in some other manner, viruses may be able to block p53 activation by directing HAUSP towards Mdm2. And because Mdm2 destabilization has such a profound effect on p53 activation, maintaining a stable Mdm2 would essentially guarantee low p53 levels and open the gate for unregulated viral replication.

The HAUSP-p53-Mdm2 Pathway

Advances in the understanding of HAUSP function have clearly placed it as a critical component of the p53-Mdm2 pathway. It was first shown to specifically deubiquitinate and stabilize p53 in both in vitro and in vivo systems (Kessler et al. 2007). Further, overexpression of HAUSP lead to p53-mediated growth suppression in p53 wildtype (H460 human large cell-lung carcinoma) but not p53 null cells (H1299 human nonsmall cell-lung carcinoma), indicating an important effect on p53 stabilization and activation. The role of HAUSP in p53 regulation became more complex with the discovery that removal of the gene in somatic HCT116 cells caused profound p53 stabilization (Cummins et al. 2004). This result was counterintuitive to what was expected, and a detailed biochemical analysis of this pathway revealed a provocative model. Two independent systems confirmed that transient partial reduction of HAUSP gene expression resulted in p53 protein destabilization, a result indicating that HAUSP does in part function as a stabilizing effector protein for p53 (Kessler et al. 2007). However, full reduction or removal of HAUSP expression leads to a destabilization of Mdm2 and dramatic stabilization of p53. HAUSP was also shown to act as a deubiquitinase for Mdm2, and this intricate interplay between p53, Mdm2, and HAUSP yields a cellular system that is delicately responsive to changes in HAUSP levels (Brooks et al. 2007). The physiology underlying this pathway remains to be elucidated. One possibility is that HAUSP is differentially regulated depending on the cellular signaling involved, so that at times it preferentially stabilizes p53 (i.e., DNA damage) and at others it stabilizes Mdm2 (i.e., normal cellular conditions) (Fig. 2). The binding affinity between HAUSP and Mdm2 is reduced after treatment with the radiomimetic agent neocarzinostatin (NCS) based on coimmunoprecipitation experiments (Meulmeester et al. 2005).

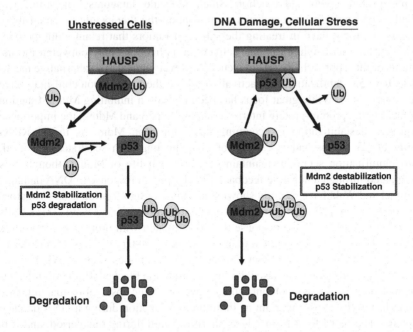

Fig. 2 A model for HAUSP interaction with p53 and Mdm2 in vivo. HAUSP preferentially binds to Mdm2 under normal unstressed conditions, leading to the deubiquitination and stabilization of this protein to ensure low levels of p53 (*left panel*). However, upon cell stress of DNA damage signals, HAUSP may lose binding ability with Mdm2 and the free molecules are available to bind to and stabilize p53, leading to protein accumulation and activation (*right panel*). Once p53 is not needed, the equilibrium would shift back to original conditions

This decreased binding was shown to be dependent on ATM-mediated phosphorylation, suggesting that phosphorylation signaling events are responsible for the decreased binding. If this signal were to occur shortly after a DNA damaging event, it would increase the pool of unbound HAUSP that would be available for p53 binding and rapid stabilization.

Though the regulation of p53 is a critical part of maintaining normal cell home-ostasis, Mdm2 can also be thought of as a critical regulatory point in this signaling pathway. Mdm2 remains the most prominent negative regulator of p53 as it is as yet the only regulator that induces p53-mediated lethality in knockout mice that is subsequently rescued when crossed with p53 knockout mice (Montes de Oca Luna et al. 1995). MdmX, another E3 ubiquitin ligase structurally similar to Mdm2, is the exception to this, as the mouse knockout of this gene behaves similarly to the Mdm2 knockout as well as the p53-Mdm2 double knockout mouse (Finch et al. 2002; Parant et al. 2001). However, the biochemical functions of MdmX are unclear as they relate to p53. Any direct regulator of Mdm2 can be thought of as an indirect mechanism for p53 regulation, and it is here where HAUSP may provide an additional regulatory layer for balancing the p53 ubiquitination pathway.

Therapeutic explorations have tended to focus on Mdm2 functions and interactions with p53. It is clear that inhibition of Mdm2 has a profound stabilizing effect on p53 protein levels and function. Small-molecule antagonists of Mdm2 can stimulate p53 both in vitro and in vivo indicating the therapeutic potential these compounds might have in treating the subset of tumors that retain wildtype p53. However, long-term inhibition and delivery complications may outweigh the net benefits of this approach. Recent evidence, however, indicates an alternative mechanism for p53 stabilization and activation whose therapeutic potential may have gone overlooked. While great focus has been placed on inhibiting Mdm2 function or blocking the protein-protein interaction between p53 and Mdm2, the importance of protein destabilization is becoming more apparent. Mdm2 is in the RING family of E3 ligases and is capable of ubiquitinating both substrates and itself. Autoubiquitination seems to maintain the short half-life of Mdm2, though it is upregulated through a negative feedback loop by p53 in response to DNA damage. Stommel and Wahl have recently shown that Mdm2 destabilization is an obligatory step for sufficient p53 stabilization and activation (Stommel and Wahl 2004). Cells undergoing a DNA damage response that were treated with proteasome inhibitors to block Mdm2 degradation failed to induce several p53 target genes. Mdm2 desta-bilization seems to involve DNA damage-induced kinases such as ATM, since treatment with the general kinase inhibitor Wortmannin blocked Mdm2 destabiliza-tion in response to stress. More intriguing, however, is the observation that inducing Mdm2 instability is as important as inhibiting its function during a DNA damage response. The half-life of Mdm2 is as short as 5 min during this period, and it is alluring to consider that phosphorylated Mdm2 may have enhanced self-ubiquitination catalysis. It is possible that the abundance of ubiquitinated Mdm2 during DNA damage is incapable of interacting with p53 due to the shear speed of degradation, as indicated by its short half-life. Additionally, Mdm2 can be degraded within the nucleus under these conditions and this may provide an additional mechanism for quick degradation.

The significant effect that DNA damage has on Mdm2 temporal stability and the rapid stabilization of p53 suggest an intimate interplay between p53, Mdm2, and HAUSP. Since the half-life of Mdm2 drastically decreases upon stimulation with DNA damage reagents and the binding affinity between HAUSP and Mdm2 goes down upon stimulation with NCS, it is quite likely that HAUSP has a direct role in these observations. Under normal, unstressed, homeostatic cell conditions, HAUSP may preferentially bind to Mdm2 and facilitate its stability at a level sufficient for maintaining p53 at low levels. It has been shown that the binding affinity between Mdm2 and HAUSP is slightly higher (K_d = 8 mM) then that between p53 and HAUSP (K_d = 10 mM) (Sheng et al. 2006). HAUSP also preferentially binds to Mdm2 in vitro when incubated with one molar equivalence of Mdm2 and tenfold molar equivalence of p53, suggesting a stronger affinity for Mdm2 when in the presence of the three proteins together (Brooks et al. 2007). Considering the fact the p53 and Mdm2 share the same binding site within the N-terminal domain of HAUSP, it is unlikely that these three proteins exist as a three protein complex at the same time. Rather, the resting state may have a shift in equilibrium in favor of HAUSP-Mdm2 complexes over HAUSP-p53 complexes (Fig. 2). This shift would ensure relatively stable Mdm2 that would maintain p53 at low levels when not needed. Upon DNA damage and cellular stresses, however, this shift would be reversed and allow for two mechanisms to take place. First, the interaction between HAUSP and Mdm2 would decrease due to stress-induced posttranslational modifications, allowing for free HAUSP molecules to interact with p53 and stabilize the protein by directly removing ubiquitin moieties. Second, the unbound Mdm2 would continue to undergo self-ubiquitination and degradation, significantly shortening its half-life and decreasing the available pool able to ubiquitinate p53. In support of this idea, it has been observed that the half-life of Mdm2 decreases drastically with a concomitant increase in p53 levels when HAUSP is removed by somatic knockout in HCT116 cells or transiently reduced in stably transfected HAUSP siRNA cells (Fig. 3). A similar response also occurs after DNA damage treatment. At an undefined point in the DNA damage response, if the DNA damage was adequately repaired to a level conducive for cell survival, posttranslational modifications would be made that restored the interaction between HAUSP and Mdm2 allowing for p53 levels to return to their resting levels.

HAUSP as an Interventional Therapeutic Target

It is well accepted that the p53 pathway is one of the most important for regulating cell growth arrest during times of DNA damage events. Without this crucial growth brake, cells containing aberrant mutations can go unchecked and propagate them from generation to generation, eventually leading to tumor growth and formation. The importance of p53 in human tumor growth was made clear with the documentation of the first 10,000 mutations found within a spectrum of tumor samples (Hainaut and Hollstein 2000). Today, p53 loss-of-function mutations are found in

a

HAUSP siRNA inducible LS88 Cell Line

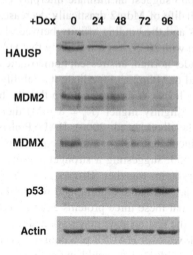

b

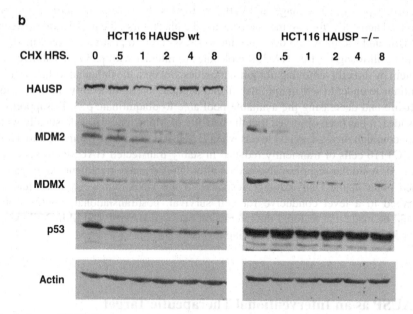

Fig. 3 Loss of HAUSP results in Mdm2 destabilization. (**a**) Reduction of HAUSP results in the destabilization of Mdm2. The HAUSP siRNA inducible stable line (LS88) was treated with doxycycline for the indicated time and Western blot analysis performed using α-HAUSP polyclonal, α-Mdm2 polyclonal, α-MdmX monoclonal, α-p53 (DO-1), and α-actin monoclonal antibodies. (**b**) The protein half-life of Mdm2 is severely compromised in the absence of HAUSP. HCT116 wt and HCT116 HAUSP$^{-/-}$ cells were treated with cyclohexamide for the indicated times and Western blot analysis performed using the above antibodies, respectively

45–50% of all human tumors taken by surgical biopsy (Caron de Fromentel and Soussi 1992; Greenblatt et al. 1994; Hollstein et al. 1991). Still, a subset of tumors retain a wildtype, functional p53 despite their uncontrolled growth, suggesting that other tumor-suppressor pathways have been altered in these cells to bypass the p53 growth arrest pathway. This also suggests that these tumors may be sensitive and responsive to cell growth arrest mechanisms mediated by p53 if it were reactivated in these cells. This, of course, has not gone overlooked, as intense focus has been placed on developing small molecule inhibitors of Mdm2 as a way of stabilizing and activating p53. The small molecule inhibitor Nutlin-3A, a cis-imidazoline analog that blocks the Mdm2-p53 interaction, has potent activation and stabilization of p53 (Vassilev et al. 2004). Nutlin-3A has also been shown to have significant success in activating wildtype p53 in several different tumor types including multiple myeloma, acute myeloid leukemia (AML), B-cell chronic lymphocytic leukemia (B-CLL), nonsmall cell lung carcinoma, neuroblastoma, malignant pleural mesothelioma, and retinoblastoma (Barbieri et al. 2006; Elison et al. 2006; Hopkins-Donaldson et al. 2006; Kojima et al. 2005; Secchiero et al. 2006; Stuhmer et al. 2005; Van Maerken et al. 2006). Targeting either Mdm2 protein–protein interactions or its ubiquitin ligase activity remains a provocative and effective strategy for stabilizing and activating p53. Still the limitations of using small molecule inhibitors through this approach may warrant alternative strategies to be investigated for two reasons. First, E3 RING ubiquitin ligases do not possess classically defined enzymatic capabilities; rather, through a somewhat undefined mechanism, they mediate the transfer of ubiquitin from an E2 conjugating enzyme to the substrate. Blocking the general function of a RING domain versus the inhibition of specific amino acids within a catalytic site proves to be much harder, particularly because the RING-mediated ubiquitin transfer is ill-defined. Second, targeting the protein–protein interaction between p53 and Mdm2 requires precise competitive inhibitors to completely block this interaction. The p53-Mdm2 contact point has not been defined down to a particular stretch of amino acids and the possibility remains that multiple interactions can occur (Brooks et al. 2007). In order to truly inhibit this interaction, a small molecule inhibitor would need to either block all possible interaction points, or several compounds would need to work in tandem for full p53 activation. Further, small molecule inhibitors such as Nutlin 3A are not active against MdmX, an increasingly important regulator of p53 function. Targeting Mdm2 exclusively may not be sufficient for potent activation of p53 in an in vivo tumor environment.

Alternatively, the data to date suggest that an equally robust p53 response can be acquired by simply blocking the catalytic capability of HAUSP. Since HAUSP is a strong regulator of both Mdm2 and MdmX, blocking the function of this protein would effectively inhibit both proteins and yield a more robust p53 response, essentially "killing two birds with one stone." The catalytic core of HAUSP is well conserved and has a precise target for inhibition, namely the cysteine-hisitidine core domain between amino acids 203 and 608. Inhibition of one amino acid,

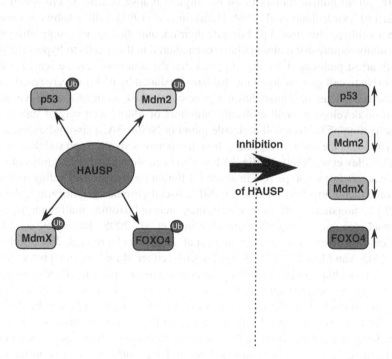

Fig. 4 HAUSP acts a specific deubiquitinating enzyme for several substrates. Schematic representations of proteins known to interact with HAUSP are indicated (on *left*). Those that are specifically deubiquitinated by HAUSP are also indicated (on *right*)

Cysteine 223, renders HAUSP completely inactive, as shown in deubiquitination assays comparing HAUSPwt to the catalytically dead mutant HAUSPcs (Kessler et al. 2007). Simply mutating one amino acid within the catalytic triad inhibits the deubiquitination activity on both Mdm2 and p53. However, due to the higher binding affinity between HAUSP and Mdm2, coupled with the observation that Mdm2 is highly destabilized in the absence of HAUSP, the strategy of blocking HAUSP activity would have the net effect of robust p53 stabilization (Fig. 4). With an increase in self-ubiquitination combined with the absence of deubiquitination, Mdm2 levels would significantly decrease allowing for the accumulation and activation of p53. Together, current data suggest that small molecule inhibitors that target the active site of HAUSP could potentially yield the same inhibition as competitive inhibitors of Mdm2. This would result in the reactivation of p53 in tumors that retain a functional copy in a much more efficient and effective way.

Conclusion

Studies on HAUSP continue to show its importance in a variety of cell processes from DNA replication to viral budding. However, its prominent role in the regulation of the p53 signaling pathway makes it an attractive target for small molecule inhibition. HAUSP deubiquitinates both MdmX and Mdm2, and as such, provides a single target for inhibiting two regulators of p53. Targeting HAUSP may therefore provide a means for activating p53 in the subset of tumors that possess a wildtype copy of the gene, effectively inhibiting two regulators with one hit. Future work will undoubtedly shed light on other physiologic roles of HAUSP as well as its precise regulatory roles in the p53 pathway.

Acknowledgments The LS174T and LS88 HAUSP siRNA cell lines were a gracious gift from M. Maurice. The HCT116 parental and HCT116 HAUSP$^{-/-}$ cell lines were a gracious gift from B. Vogelstein. The α-Mdm2 polyclonal antibody was raised against the N-terminus of Mdm2 (GST-Mdm2 residues 1–110). The α-HAUSP polyclonal antibody was raised against the N-terminus of HAUSP (GST-HAUSP residues 1–67). Both polyclonal antibodies were further affinity-purified on the antigen column. The MdmX antibody (8C6) was a generous gift from J. Chen. This work was supported in part by grants from NIH/NCI to W.G.

References

Amerik, A. Y., and Hochstrasser, M. (2004). Mechanism and function of deubiquitinating enzymes. Biochim Biophys Acta 1695, 189–207.

Barbieri, E., Mehta, P., Chen, Z., Zhang, L., Slack, A., Berg, S., and Shohet, J. M. (2006). MDM2 inhibition sensitizes neuroblastoma to chemotherapy-induced apoptotic cell death. Mol Cancer Ther 5, 2358–2365.

Boutell, C., Canning, M., Orr, A., and Everett, R. D. (2005). Reciprocal activities between herpes simplex virus type 1 regulatory protein ICP0, a ubiquitin E3 ligase, and ubiquitin-specific protease USP7. J Virol 79, 12342–12354.

Brooks, C. L., Li, M., Hu, M., Shi, Y., and Gu, W. (2007). The p53-Mdm2-HAUSP complex is involved in p53 stabilization by HAUSP. Oncogene 26, 7262–7266.

Caron de Fromentel, C., and Soussi, T. (1992). TP53 tumor suppressor gene: a model for investigating human mutagenesis. Genes Chromosomes Cancer 4, 1–15.

Cummins, J. M., Rago, C., Kohli, M., Kinzler, K.W., Lengauer, C., and Vogelstein, B., (2004). Tumour suppression: disruption of HAUSP gene stabilizes p53. Nature 428, 1p following 486.

Elison, J. R., Cobrinik, D., Claros, N., Abramson, D. H., and Lee, T. C. (2006). Small molecule inhibition of HDM2 leads to p53-mediated cell death in retinoblastoma cells. Arch Ophthalmol 124, 1269–1275.

Everett, R. D., Meredith, M., and Orr, A. (1999). The ability of herpes simplex virus type 1 immediate-early protein Vmw110 to bind to a ubiquitin-specific protease contributes to its roles in the activation of gene expression and stimulation of virus replication. J Virol 73, 417–426.

Everett, R. D., Meredith, M., Orr, A., Cross, A., Kathoria, M., and Parkinson, J. (1997). A novel ubiquitin-specific protease is dynamically associated with the PML nuclear domain and binds to a herpesvirus regulatory protein. EMBO J 16, 1519–1530.

Finch, R. A., Donoviel, D. B., Potter, D., Shi, M., Fan, A., Freed, D. D., Wang, C. Y., Zambrowicz, B. P., Ramirez-Solis, R., Sands, A. T., and Zhang, N. (2002). mdmx is a negative regulator of p53 activity in vivo. Cancer Res 62, 3221–3225.

Greenblatt, M. S., Bennett, W. P., Hollstein, M., and Harris, C. C. (1994). Mutations in the p53 tumor suppressor gene: clues to cancer etiology and molecular pathogenesis. Cancer Res 54, 4855–4878.

Hainaut, P., and Hollstein, M. (2000). p53 and human cancer: the first ten thousand mutations. Adv Cancer Res 77, 81–137.

Hershko, A., and Ciechanover, A. (1998). The ubiquitin system. Annu Rev Biochem 67, 425–479.

Hollstein, M., Sidransky, D., Vogelstein, B., and Harris, C. C. (1991). p53 mutations in human cancers. Science 253, 49–53.

Holowaty, M. N., Zeghouf, M., Wu, H., Tellam, J., Athanasopoulos, V., Greenblatt, J., and Frappier, L. (2003). Protein profiling with Epstein-Barr nuclear antigen-1 reveals an interaction with the herpesvirus-associated ubiquitin-specific protease HAUSP/USP7. J Biol Chem 278, 29987–29994.

Hong, S., Kim, S. J., Ka, S., Choi, I., and Kang, S. (2002). USP7, a ubiquitin-specific protease, interacts with ataxin-1, the SCA1 gene product. Mol Cell Neurosci 20, 298–306.

Hopkins-Donaldson, S., Belyanskaya, L. L., Simoes-Wust, A. P., Sigrist, B., Kurtz, S., Zangemeister-Wittke, U., and Stahel, R. (2006). p53-induced apoptosis occurs in the absence of p14(ARF) in malignant pleural mesothelioma. Neoplasia 8, 551–559.

Kessler, B. M., Fortunati, E., Melis, M., Pals, C. E., Clevers, H., and Maurice, M. M. (2007). Proteome changes induced by knock-down of the deubiquitylating enzyme HAUSP/USP7. J Proteome Res 6, 4163–4172.

Kojima, K., Konopleva, M., Samudio, I. J., Shikami, M., Cabreira-Hansen, M., McQueen, T., Ruvolo, V., Tsao, T., Zeng, Z., Vassilev, L. T., and Andreeff, M. (2005). MDM2 antagonists induce p53-dependent apoptosis in AML: implications for leukemia therapy. Blood 106, 3150–3159.

Laney, J. D., and Hochstrasser, M. (1999). Substrate targeting in the ubiquitin system. Cell 97, 427–430.

Lee, H. J., Kim, M. S., Kim, Y. K., Oh, Y. K., and Baek, K. H. (2005). HAUSP, a deubiquitinating enzyme for p53, is polyubiquitinated, polyneddylated, and dimerized. FEBS Lett 579, 4867–4872.

Meulmeester, E., Maurice, M. M., Boutell, C., Teunisse, A. F., Ovaa, H., Abraham, T. E., Dirks, R. W., and Jochemsen, A. G. (2005). Loss of HAUSP-mediated deubiquitination contributes to DNA damage-induced destabilization of Hdmx and Hdm2. Mol Cell 18, 565–576.

Millard, S. M., and Wood, S. A. (2006). Riding the DUBway: regulation of protein trafficking by deubiquitylating enzymes. J Cell Biol 173, 463–468.

Montes de Oca Luna, R., Wagner, D. S., and Lozano, G. (1995). Rescue of early embryonic lethality in mdm2-deficient mice by deletion of p53. Nature 378, 203–206.

Nijman, S. M., Luna-Vargas, M. P., Velds, A., Brummelkamp, T. R., Dirac, A. M., Sixma, T. K., and Bernards, R. (2005). A genomic and functional inventory of deubiquitinating enzymes. Cell 123, 773–786.

Parant, J., Chavez-Reyes, A., Little, N. A., Yan, W., Reinke, V., Jochemsen, A.G., and Lozano, G. (2001). Rescue of embryonic lethality in Mdm4-null mice by loss of Trp53 suggests a nonoverlapping pathway with MDM2 to regulate p53. Nat Genet 29, 92–95.

Pickart, C.M., and Eddins, M.J. (2004). Ubiquitin: structures, functions, mechanisms. Biochim Biophys Acta 1695, 55–72.

Puente, X.S., and Lopez-Otin, C. (2004). A genomic analysis of rat proteases and protease inhibitors. Genome Res 14, 609–622.

Robinson, P.A., Lomonte, P., Leek Markham, A.F., and Everett, R.D. (1998). Assignment1 of herpesvirus-associated ubiquitin-specific protease gene HAUSP to human chromosome band 16p13.3 by in situ hybridization. Cytogenet Cell Genet 83, 100.

Secchiero, P., Barbarotto, E., Tiribelli, M., Zerbinati, C., di Iasio, M.G., Gonelli, A., Cavazzini, F., Campioni, D., Fanin, R., Cuneo, A., and Zauli, G. (2006). Functional integrity of the p53-mediated apoptotic pathway induced by the nongenotoxic agent nutlin-3 in B-cell chronic lymphocytic leukemia (B-CLL). Blood 107, 4122–4129.

Sheng, Y., Saridakis, V., Sarkari, F., Duan, S., Wu, T., Arrowsmith, C.H., and Frappier, L. (2006). Molecular recognition of p53 and MDM2 by USP7/HAUSP. Nat Struct Mol Biol 13, 285–291.

Stommel, J.M., and Wahl, G.M. (2004). Accelerated MDM2 auto-degradation induced by DNA-damage kinases is required for p53 activation. EMBO J 23, 1547–1556.

Stuhmer, T., Chatterjee, M., Hildebrandt, M., Herrmann, P., Gollasch, H., Gerecke, C., Theurich, S., Cigliano, L., Manz, R.A., Daniel, P.T., et al. (2005). Nongenotoxic activation of the p53 pathway as a therapeutic strategy for multiple myeloma. Blood 106, 3609–3617.

Tang, J., Qu, L.K., Zhang, J., Wang, W., Michaelson, J.S., Degenhardt, Y.Y., El-Deiry, W.S., and Yang, X. (2006). Critical role for Daxx in regulating Mdm2. Nat Cell Biol 8, 855–862.

van der Horst, A., de Vries-Smits, A.M., Brenkman, A.B., van Triest, M.H., van den Broek, N., Colland, F., Maurice, M.M., and Burgering, B.M. (2006). FOXO4 transcriptional activity is regulated by monoubiquitination and USP7/HAUSP. Nat Cell Biol 8, 1064–1073.

van der Knaap, J.A., Kumar, B.R., Moshkin, Y.M., Langenberg, K., Krijgsveld, J., Heck, A.J., Karch, F., and Verrijzer, C.P. (2005). GMP synthetase stimulates histone H2B deubiquitylation by the epigenetic silencer USP7. Mol Cell 17, 695–707.

Van Maerken, T., Speleman, F., Vermeulen, J., Lambertz, I., De Clercq, S., De Smet, E., Yigit, N., Coppens, V., Philippe, J., De Paepe, A., et al. (2006). Small-molecule MDM2 antagonists as a new therapy concept for neuroblastoma. Cancer Res 66, 9646–9655.

Vassilev, L.T., Vu, B.T., Graves, B., Carvajal, D., Podlaski, F., Filipovic, Z., Kong, N., Kammlott, U., Lukacs, C., Klein, C., et al. (2004). In vivo activation of the p53 pathway by small-molecule antagonists of MDM2. Science 303, 844–848.

Modulation of Protein Stability: Targeting the VHL Pathway

William Y. Kim and William G. Kaelin, Jr.

Abstract Inactivation of the von Hippel–Lindau tumor suppressor protein (pVHL) has been linked to a variety of tumors such as renal cell carcinoma, pheochromocytomas, and cerebellar hemangioblastomas. The best characterized of its many proposed functions is the ability to downregulate hypoxia-inducible factor-α (HIFα) subunits. Inactivation of pVHL and its ubiquitin ligase activity result in the stabilization of HIFα subunits and the transactivation of HIF target genes. Therapeutic approaches that restore pVHL action or inhibit HIF transcriptional activity remain unrealized and will likely remain a therapeutic challenge. Several pVHL independent pathways that alter HIFα protein stability have recently been identified, and the availability of inhibitors in clinical trials makes these pathways attractive for further investigation.

Keywords von Hippel–Lindau • Hypoxia-inducible factor • mTOR • HSP-90 • Ubiquitination

von Hippel–Lindau Disease

von Hippel–Lindau (VHL) disease is a familial cancer syndrome first described in the medical literature over 100 years ago. VHL patients are predisposed to developing a spectrum of benign and malignant tumors. Affected individuals frequently develop clear cell renal cell carcinomas, pheochromocytomas, central nervous system hemangioblastomas, retinal angiomas, and cysts of visceral organs (typically, the pancreas and kidney). Other manifestations of VHL disease include pancreatic islet cell tumors (classically biochemically inactive), endolymphatic sac tumors, and

W.Y. Kim (✉) and W.G. Kaelin, Jr. (✉)
Department of Medicine, Division of Hematology/Oncology,
University of North Carolina, Chapel Hill, NC, USA
e-mail: wykim@med.unc.edu

K. Sakamoto and E. Rubin (eds.), *Modulation of Protein Stability in Cancer Therapy*,   45
DOI: 10.1007/978-0-387-69147-3_4, © Springer Science+Business Media, LLC 2009

cystadenomas of the pancreas and epididymis (Kim and Kaelin 2004). Patients with VHL disease inherit a defective allele of the tumor suppressor gene, *VHL*. The development of pathology is linked to somatic inactivation of the remaining wild-type *VHL* allele, resulting in loss of function of the *VHL* gene product, pVHL. Since the probability of this event occurring in at least one susceptible cell during a carrier's lifetime is very high (~90%), VHL disease appears to be transmitted in an autosomal dominant manner even though it is caused by a recessive mutation. Thus, the *VHL* gene is a classic 2-Hit tumor suppressor gene (Knudson 1971).

The VHL Protein, pVHL

The *VHL* gene was isolated by positional cloning in 1993, after earlier genetic linkage studies had indicated that it was located on chromosome 3p25 (Seizinger et al. 1988; Latif et al. 1993). The *VHL* gene is composed of three exons and its mRNA is ubiquitously expressed. Significantly, *VHL* gene expression is not restricted to the tissues that give rise to VHL-associated tumors. Therefore, the pattern of *VHL* expression does not account for the differences in organ-specific cancer risk (Kaelin 2002). The full-length mRNA of *VHL* contains two different, in-frame, ATG codons (codons 1 and 54), each of which can serve as translational initiation sites, thus resulting in two different protein products (often referred to as pVHL19 and pVHL30 based on their apparent molecular masses upon protein electrophoresis) (Schoenfeld et al. 1998). Early studies showed that the two proteins behaved similarly in most biochemical and functional assays and thus they are often referred to generically as pVHL (Iliopoulos et al. 1998; Blankenship et al. 1999). However, recent data indicate important differences between the two gene products. For example, both kinases casein kinase 2 (CK2) and GSK3β phosphorylate only pVHL30 (Lolkema et al. 2005; Hergovich et al. 2006). CK2 phosphorylates pVHL30 on its N-terminus and this modification is necessary for pVHL30's ability to appropriately bind and properly direct the deposition of fibronectin in the extracellular matrix (Lolkema et al. 2005). GSK3β phosphorylates pVHL30 on serine 68 activating its microtubule-stabilizing capacity (Hergovich et al. 2003).

pVHL30 and pVHL19 appear to have different patterns of subcellular localization as well. The pVHL proteins produced upon introduction of the cDNA encoding pVHL30 (which also results in the expression of pVHL19) into *VHL−/−* cells have been reported to reside primarily in the cytoplasm but can also be found in the nucleus, in mitochondria, and in association with the endoplasmic reticulum. However, more recent studies using pVHL19- and pVHL30-specific antibodies have shown that the majority of endogenous pVHL30 and pVHL19 reside in the cytoplasm and nucleus, respectively (Hergovich et al. 2003). Other reports have demonstrated that pVHL shuttles dynamically back and forth between the nucleus and cytoplasm and cannot suppress tumor growth when artificially restrained from doing so (Lee et al. 1996, 1999).

VHL Polyubiquitinates HIF

Early biochemical studies revealed that pVHL formed stable complexes with Elongin B and Elongin C (Duan et al. 1995; Kibel et al. 1995; Kishida et al. 1995). Soon thereafter, two additional components of the pVHL complex, called Cul2 and Rbx1, were discovered (Pause et al. 1997; Lonergan et al. 1998; Kamura et al. 1999). Elongin C and Cul2 were noted to have structural homology to the Skp1 and Cdc53 components of SCF (Skp1, Cdc53, F-Box protein) ubiquitin ligases found in yeast (Bai et al. 1996; Pause et al. 1997; Lonergan et al. 1998). The three-dimensional crystal structure of pVHL bound to Elongin B and Elongin C confirmed that Elongin C, as suspected, resembled Skp1 and revealed that the region of pVHL that binds directly to Elongin C loosely resembles an F-box (Stebbins et al. 1999). Therefore, the hypothesis emerged that the pVHL complex, like SCF complexes in yeast, might act as an ubiquitin ligase.

Overproduction of hypoxia-inducible mRNAs, irrespective of oxygen availability, is a hallmark of pVHL-defective tumor cells (Iliopoulos et al. 1996). In 1999, Maxwell and coworkers showed that cells lacking pVHL failed to degrade the α subunit of the transcription factor, hypoxia-inducible factor-1 (HIF1) in the presence of oxygen and that pVHL bound to HIF (Maxwell et al. 1999). Soon thereafter, it was shown that the pVHL complex was indeed a ubiquitin ligase and that in this context, pVHL, similar to the role played by F-box proteins in yeast, acts as the substrate recognition module that brings the ubiquitin conjugating machinery into the proximity of its substrate, HIF (Cockman et al. 2000; Kamura et al. 2000; Ohh et al. 2000; Tanimoto et al. 2000).

HIF is a heterodimeric transcription factor, consisting of a labile α subunit and a stable β subunit. The human genome contains three HIF-α genes (HIF1α, HIF2α, and HIF3α) as well as three HIF-β genes, also called the arylhydrocarbon receptor nuclear translocators (ARNT) (Semenza 2003). Under low oxygen conditions or in cells lacking pVHL, HIFα accumulates, binds to HIFβ, and transcriptionally activates genes whose promoters contain specific DNA sequences called hypoxia-response elements (HREs). Many of the genes regulated by HIF promote adaptation to acute or chronic hypoxia. These include genes implicated in the uptake and metabolism of glucose, apoptosis, angiogenesis, autophagy, control of extracellular pH, mitogenesis, and erythropoiesis (Semenza 2003).

The stabilization of HIFα that normally occurs under hypoxic conditions implied that the binding of pVHL to HIFα was oxygen-dependent. Earlier studies identified a region of HIF1α, sometimes called the ODD (oxygen-dependent degradation domain), which was sufficient to confer instability in the presence of oxygen (Huang et al. 1998). It was later shown that when oxygen is plentiful, HIFα subunits become hydroxylated on one (or both) of two conserved prolyl residues within the ODD (Ivan et al. 2001; Jaakkola et al. 2001; Masson et al. 2001; Yu et al. 2001) by members of the iron- and 2-oxogluterate-dependent EglN prolyl hydroxylase family (Bruick and McKnight 2001; Epstein et al. 2001; Ivan et al. 2002). Hydroxylation of either of the prolyl residues creates a high-affinity pVHL-binding site. The recruitment of pVHL then leads to HIFα polyubiquitylation and destruction via the 26S proteasome (Ivan et al. 2001; Jaakkola et al. 2001; Masson et al. 2001; Yu et al. 2001). Conversely, under low-oxygen conditions, HIFα is not hydroxylated and thus escapes recognition by pVHL (Fig. 1).

pVHL contains two mutational hotspots called the alpha domain and the beta domain. The alpha domain is necessary to bind elongin C through which pVHL forms stable complexes with the other members of the pVHL complex (Kaelin 2002). The beta domain contains sequences necessary for substrate recognition and thus binds hydroxylated HIF and possibly other, as yet to be determined, pVHL substrates. Renal carcinoma-associated pVHL mutants are invariably defective with respect to HIF polyubiquitination, suggesting that HIF, or perhaps some other beta-domain binding protein, plays a critical role in renal cell carcinogenesis (Kaelin 2002).

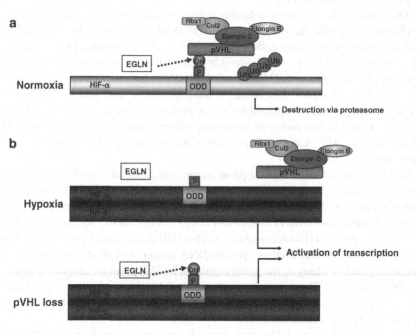

Fig. 1 Regulation of HIF by the pVHL complex. (**a**) Under normal oxygen conditions, HIFα subunits are hydroxylated on conserved prolyl residues located within the oxygen-dependent degradation domain (ODD). This post-translational modification is catalyzed by members of the EglN prolyl hydroxylase family. Prolyl hydroxylation of HIFα subunits results in the recruitment of the pVHL complex composed of Elongin B, Elongin C, Cul2, and Rbx1. pVHL, in turn, orchestrates the polyubiquitination of HIFα marking it for destruction via proteasome. (**b**) Under hypoxic conditions, EglN is inactive and consequently HIF is not hydroxylated. Alternatively, in the absence of pVHL, while HIFα is hydroxylated it is not polyubiquitinated. In both cases, HIFα is stabilized, heterodimerizes with HIFβ, and activates the transcription of genes containing hypoxia response elements (HREs)

The VHL Pathway in Renal Cell Carcinoma

In the context of VHL disease, inactivation of the *VHL* gene is an early step in clear cell renal carcinogenesis (Mandriota et al. 2002). *VHL* inactivation is also common in nonhereditary clear cell renal carcinoma, as approximately 50% of sporadic clear cell renal carcinomas harbor (somatic) mutations affecting the maternal and paternal *VHL* locus. Of the remaining 50% of tumors, perhaps one in five will fail to transcribe the *VHL* gene because of promoter hypermethylation (Kim and Kaelin 2004). Therefore, the majority of clear cell renal carcinomas appear to be linked to biallelic *VHL* inactivation. It is unlikely, however, that loss of *VHL* is sufficient for renal tumorigenesis. Consistent with this notion, conditional inactivation of *VHL* in mouse kidney proximal tubular cells (postulated to be the precursor cells to renal cell carcinoma) causes the development of renal cysts but not renal cell carcinoma (Rankin et al. 2006). Therefore, *VHL* appears to act as a gatekeeper tumor suppressor gene for human clear cell renal carcinoma. Uncovering the mutations that cooperate with *VHL* loss to promote renal carcinogenesis is an area of active investigation.

There is mounting evidence that HIFα, and especially HIF2α, plays a causal role in the development of pVHL-defective renal carcinomas and is not merely a marker for disruption of the pVHL pathway. Examination of the kidneys of VHL patients reveals early preneoplastic lesions with increased accumulation of HIF1α. Intriguingly, there is an apparent switch from HIF1α accumulation to HIF2α accumulation in such lesions that is coincident with increasing dysplasia, raising the possibility that HIF2α is more oncogenic than HIF1α (Mandriota et al. 2002). Other lines of evidence support this position as well. First, human renal carcinoma lines typically either express both HIF1α and HIF2α together or express HIF2α alone. This suggests that there is a selection pressure to maintain HIF2α expression and, possibly, selection pressure to diminish HIF1α expression (Maxwell et al. 1999). Second, restoring the function of pVHL in VHL−/− renal carcinoma cells prevents them from forming tumors after injection into nude mice (Iliopoulos et al. 1995). This action of pVHL, however, is neutralized in cells engineered to produce a HIF2α variant that escapes recognition by pVHL (because their proline hydroxylation sites have been eliminated), but not the corresponding HIF1α variant (Iliopoulos et al. 1995; Kondo et al. 2002, 2003; Maranchie et al. 2002). This suggests that inhibition of HIFα, and especially HIF2α, is *necessary* for tumor suppression by pVHL. Finally, silencing of HIF2α by retroviruses encoding short-hairpin RNAs against HIF2α in human *VHL*−/− human renal carcinomas is sufficient to prevent them from forming tumors in nude mice (Kondo et al. 2003; Zimmer et al. 2004). In mice, inactivation or stabilization of HIF prevents or causes, respectively, many of the pathological changes seen in the setting of pVHL loss, implying that HIF is necessary and sufficient for these phenotypes (Rankin et al. 2005, 2006; Kim et al. 2006). These experiments "validate" HIFα as a therapeutic target in clear cell renal carcinoma and provide a rationale for the development of either direct HIF antagonists or inhibitors of proteins that contribute to tumor growth downstream of HIF.

Targeting HIF Protein Stability

The most intellectually satisfying approach to the therapy of pVHL-defective tumors would be to restore the missing wild-type protein by using methods such as "gene therapy." Unfortunately, replacement of faulty or absent genes by gene therapy has been plagued with unacceptable toxicities including malignancies secondary to insertional mutagenesis by retroviral vectors (Hacein-Bey-Abina et al. 2003). Attempts on gene therapy with the *VHL* gene have not been reported in humans. However, an adenovirus encoding the human form of *VHL*, injected into the preretinal space, has been used to inhibit retinal neoangiogenesis in monkeys (Akiyama et al. 2004).

An alternative approach would be to identify drug-like molecules that inhibit the HIF transcription factor. Unfortunately, inhibiting the transcriptional activity of transcription factors, by blocking their binding to co-transcription factors, co-activators, or directly to DNA with drug-like small organic molecules, has generally been difficult, with the exception of the steroid hormone receptors. Modulating HIFα protein levels, however, appears more promising.

mTOR Inhibitors

While the predominant determinant of HIFα protein abundance is the presence of an intact pVHL pathway, there are pVHL-independent determinants of HIFα protein stability as well. HIFα subunits have a very short half-life (measured in minutes during normoxia) and therefore, similar to other proteins with short half-lives, are sensitive to changes in transcription or protein translation, such as those mediated by the mammalian target of rapamycin (mTOR).

mTOR is a serine/threonine kinase that functions as a master regulator of protein translation and a gatekeeper of cell growth, metabolism, and proliferation (Shaw and Cantley 2006). mTOR, in association with the regulatory associated protein of TOR (Raptor) and GβL, comprise the mTORC1 complex. mTORC1 phosphorylates two key substrates, S6Kinase (S6K) and 4E-BP1 (Thomas 2006). Phosphorylation and activation of S6K results in increased translation of mRNAs containing conserved sequences within their 5′ untranslated regions, so called TOP (terminal oligopyrimidine tracts) sequences, including mRNAs that encode many components of the translational machinery (Meyuhas 2000). Additionally, mTORC1 indirectly regulates the eukaryotic initiation factor 4E (eIF-4E) by phosphorylating its inhibitory partner 4E-BP1 resulting in the release of eIF-4E and the initiation of protein translation of mRNAs with regulatory elements in the 5′ untranslated terminal regions, so called CAP-dependent mRNAs, such as the mRNAs encoding cyclin D1 and c-MYC (Thomas 2006; Rosenwald et al. 1995; Mendez et al. 1996). In total, mTORC1's kinase activity promotes an increase in global translation. Accordingly, mTORC1 inhibitors downregulate HIFα protein levels (Fig. 2a) (Hudson et al. 2002;

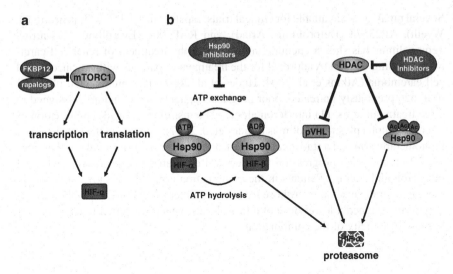

Fig. 2 Modulation of HIF protein stability. (**a**) The mTORC1 complex regulates the translation of numerous mRNAs, including the HIF1α mRNA, as well as HIF1α transcription. Rapamycin and its analogs (rapalogs) bind to the endogenous, intracellular, protein called FKBP12. The FKBP12/rapamycin complex binds to, and inhibits, the mTORC1 kinase resulting in decreased HIF transcription and translation. (**b**) ATP-associated Hsp90 aids in the folding and stabilization of HIF1α and other so-called client proteins while ADP-associated Hsp90 promotes the ubiquitination and proteasome-mediated degradation of these same proteins. Inhibitors of Hsp90 block its activity by competitively binding to its nucleotide pocket and mimicking ADP-associated Hsp90. Inhibition of HDAC activity results in the pVHL-dependent or pVHL-independent destruction of HIF via the proteasome. HDAC inhibition results in the upregulation of pVHL expression and secondary downregulation of HIF. Alternatively, HDAC inhibitors promote the hyperacetylation and inactivation of Hsp90 triggering the destabilization of its client proteins

Treins et al. 2002; Brugarolas et al. 2003; Thomas et al. 2006). Moreover, in addition to regulating HIFα translation, mTORC1 inhibitors also downregulate HIF1α transcription (Brugarolas et al. 2003).

Preclinical studies show that the treatment of pVHL-defective cells with compounds that inhibit mTORC1 activity results in a decrease in HIFα protein levels and growth inhibition of *VHL*-null renal carcinoma xenografts (Hudson et al. 2002; Treins et al. 2002; Brugarolas et al. 2003; Thomas et al. 2006). This effect is partially mediated by the presence of TOP tracts located in the 5'UTRs of both HIF1 and HIF2 (Thomas et al. 2006) as protein levels of HIF variants lacking their native 5' UTRs (and thus their TOP tracts) are insensitive to mTOR inhibitors and their growth inhibitory effects (Thomas et al. 2006).

Rapamycin and rapamycin-like inhibitors of mTORC1 ("rapalogs") mediate their effect by complexing with the endogenous, intracellular cofactor, FKBP12 (Fig. 2a). The resulting complex inhibits mTORC1 and subsequent translational initiation.

Several rapalogs are available for clinical trials: rapamycin, CCI-779 (Temsirolimus, Wyeth), AP23573 (Deforolimus, Ariad), and RAD-001 (Everolimus, Novartis). Temsirolimus has shown encouraging results in the treatment of renal cell carcinoma, leading to its FDA approval for the treatment of patients with advanced renal cell carcinoma (Atkins et al. 2004; Hudes et al. 2007). In a multicenter, phase III trial, 626 previously untreated, poor prognosis patients were randomly assigned to three treatment arms: (1) interferon alpha, (2) temsirolimus, or (3) combination of interferon alpha plus temsirolimus (Hudes et al. 2007). Patients in the temsirolimus monotherapy arm had a statistically significant longer overall survival (10.9 months vs. 7.3 months) and progression-free survival compared to patients who received interferon alpha alone. Patients in the combination arm did not show an improvement in overall survival over those treated with interferon alone, possibly because they were assigned a lower dose of temsirolimus, which was often reduced further because of toxicity of the combination.

Hsp90 Inhibitors

The molecular chaperone, heat shock protein 90 (Hsp90), plays an essential role in facilitating the proper folding, stabilization, and localization of its so-called client proteins. Hsp90 possesses intrinsic ATPase activity and upon binding ATP, hydrolyzes it to ADP. Thus, Hsp90 cycles between an ATP or ADP-bound state and this continuous cycling is critical for its chaperone functions (Fig. 2b) (Isaacs et al. 2003). ATP and ADP-bound forms of Hsp90 associate with distinct co-chaperone proteins. ATP-associated Hsp90 results in the recruitment of co-chaperones that aide in client protein folding and stabilization while ADP-associated Hsp90 promotes the ubiquitination and proteasome-mediated degradation of these same proteins. Compounds such as geldenamycin and the small molecule analog of geldenamycin, 17-allylamino-17-desmethoxygeldanamycin (17-AAG), block Hsp90 activity by competitively binding to its nucleotide pocket, mimicking ADP-associated Hsp90 (Isaacs et al. 2003).

HIF1α is an Hsp90 client protein and as predicted, inhibition of Hsp90 downregulates HIFα protein levels (Gradin et al. 1996; Isaacs et al. 2002; Mabjeesh et al. 2002). In this scenario, HIFα is polyubuitinated and degraded by a "pVHL-like" complex that is identical to the pVHL complex with the exception that the receptor of activated protein kinase C (RACK1) substitutes for pVHL (Liu et al. 2007). Importantly, the sensitivity of HIFα to Hsp90 inhibition occurs in cells that lack functional pVHL, suggesting that Hsp90 inhibitors may be active in VHL-null renal cell carcinomas (Isaacs et al. 2002). Consistent with this notion, some phase 1 and 2 clinical trials using 17AAG have included patients with renal cell carcinoma (both clear cell and papillary histologies) and these patients have demonstrated variable lengths of disease stabilization (Ronnen et al. 2006; Solit et al. 2007; Vaishampayan et al. 2007).

HDAC Inhibitors

There is increasing recognition that epigenetic changes to chromatin play a role in the regulation of gene expression within tumor cells. A range of post-translational modifications to the amino terminus of histones have now been uncovered, including acetylation, methylation, phosphorylation, ubiquitination, sumolyation, and glycosylation (Bolden et al. 2006). Histone acetyltransferases (HATs) and histone deacetylases (HDACs) have opposing effects on gene expression. Whereas HATs transfer acetyl groups to histones resulting in euchromatinization of DNA and the increased accessibility of transcriptional machinery to promoters of genes, HDACs catalyze the removal of acetyl groups resulting in heterochromatinization and transcriptional repression.

HDAC inhibitors can directly induce cancer cell death. However, their indirect effects on angiogenesis have been noted only recently (Bolden et al. 2006). Pharmacologic inhibition of HDAC activity results in a decrease in HIFα protein levels (Fig. 2b) (Kim et al. 2001; Jeong et al. 2002; Kong et al. 2006; Qian et al. 2006). Early reports suggested that HDAC inhibitors mediate their antiangiogenic effects via upregulation of pVHL expression and secondary downregulation of HIF (Kim et al. 2001). However, subsequent studies have demonstrated that HDAC inhibition results in the proteasomal destruction of HIF in a pVHL-independent manner through the promotion of Hsp90 hyperacetylation and inactivation (Kong et al. 2006; Qian et al. 2006). In this regard, HDAC inhibitors have been shown to act synergistically with Hsp90 inhibitors to degrade the chronic myelogenous leukemia (CML) associated BCR-ABL fusion protein as well as mutant forms of FLT3 implicated in acute myelogenous leukemia (AML) (Rahmani et al. 2003; George et al. 2004). This combination has not yet been evaluated for its effects on HIF degradation or in the context of renal cell carcinoma. A number of other drugs have been identified that indirectly downregulate HIFα protein levels including thioredoxin inhibitors, some topiosomerase I inhibitors, and certain inhibitors of microtubule polymerization (Kaelin 2005; Melillo 2007). While most of these agents appear to affect HIF protein translation, their precise mechanism of action remains unclear (Mabjeesh et al. 2003; Rapisarda et al. 2004).

Targeting HIF-Responsive Growth Factors and Growth Factor Receptors

More than 100 genes that are direct transcriptional targets of HIF have been described with a number of these genes encoding proteins that are growth factor receptors or their ligands (Caldwell et al. 2002; Galban et al. 2003; Jiang et al. 2003; Semenza 2003). Disruption of several of these growth factor pathways by blocking either antibodies or small-molecule inhibitors has shown therapeutic efficacy in clinical trials (Fig. 3).

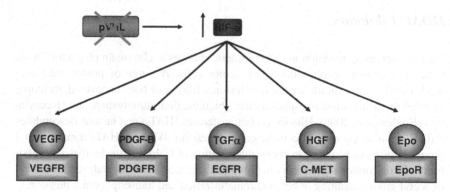

Fig. 3 Targeting HIF-responsive growth factors. Loss of pVHL and consequent HIF upregulation results in the transactivation of HIF-responsive growth factors and/or their cognate receptors, many of which are transmembrane receptor tyrosine kinases (RTKs). Therapeutic inhibition of these pathways by small organic molecules that bind within the ATP pocket of RTKs or by humanized, inhibitory, monoclonal antibodies have shown mixed success in kidney cancer clinical trials. The most consistent beneficial effects thus far have been seen with agents directed against VEGF or its receptor, KDR

Vascular Endothelial Growth Factor

Clinically, clear cell renal carcinomas are known to be highly vascular tumors. In the laboratory, renal carcinoma cell lines and patient-derived tumors have been found to have high levels of a variety of pro-angiogenic molecules including the HIF-responsive gene product vascular endothelial growth factor (VEGF). VEGF stimulates endothelial cell proliferation and survival. Accordingly, agents which inhibit the VEGFR are thought to work primarily by inhibiting angiogenesis. However, in one study, the VEGF receptor, kinase insert domain-containing receptor (KDR), has been detected on renal carcinoma cells, suggesting the possibility that it may act as an autocrine growth factor as well (Fox et al. 2004). This observation awaits confirmation.

Several drugs which bind soluble VEGF, or inhibit KDR have shown clinical activity against renal cell carcinoma. In a randomized phase II study, patients with metastatic renal carcinoma who were treated with a neutralizing antibody against VEGF, bevacizumab (Avastin, Genetech), exhibited a significant delay in time-to-disease progression (Yang et al. 2003). This activity has now been confirmed in a randomized Phase III trial in which patients receiving alpha-interferon were randomized to also receive bevacizumab or placebo (Escudier et al. 2007). In addition, multiple structurally unrelated KDR inhibitors, including sunitinib (Sutent, Pfizer), sorafenib (Nexavar, Bayer), GW786034 (Pazopanib, GlaxoSmithKline), and AG013736 (axitinib, Pfizer) also appear to have significant activity against this tumor subtype (Escudier et al. 2007; Hutson et al. 2007; Motzer et al. 2007; Rixe et al. 2007).

Two of these compounds, sorafenib and sunitinib, are now FDA approved for the treatment of advanced renal cell carcinoma. Sorafenib was initially developed as an inhibitor of the B-Raf and C-Raf kinases. However, in early trials, its activity in kidney cancer, relative to other epithelial neoplasms, prompted the discovery that it inhibits KDR as well as other kinases such as VEGFR-3, and platelet-derived growth factor receptor (PDGFR). It is possible that RAF inhibition by sorafenib is, however, relevant in this setting since RAF has been implicated in endothelial cell survival (Hood et al. 2002). A placebo-controlled phase III trial of patients with cytokine refractory meta-static renal cell carcinoma showed a statistically significant improvement in median progression-free survival of 5.5 months for patients on sorafenib vs. 2.8 months for patients treated with placebo (Escudier et al. 2007). After an early progression-free survival analysis, a decision was made to administer sorafenib to patients who were assigned to receive placebo; this occurred in approximately 50% of the patients on the placebo arm. Ultimately, in an intention to treat analysis no statistically significant improvement in overall survival was observed. Whether this was secondary to the high rate of patient cross-over from placebo to sorafenib cannot be determined.

Sunitinib, another small-molecule multikinase inhibitor blocks the activity of KDR as well as PDGFR, insulin-like growth factor receptor 1 (IGFR-1), c-Src, and c-Abl. (Costa and Drabkin 2007). A multi-institutional phase III trial randomized 750 previously untreated patients with clear cell renal carcinoma to receive sunitinib or interferon (IFN) (Motzer et al. 2007). Patients treated with sunitinib experienced a higher objective response rate (31% vs. 6%), a longer progression free survival (11 months vs. 5 months), as well as a better quality of life.

Ultimately, however, tumors become resistant to these multikinase inhibitors. While there is significant cross-resistance between the sunitinib and sorafenib, there is now clinical data to support the use of sunitinib in sorafenib failures and vice versa (Dham and Dudek 2007). An important outstanding clinical question is whether the apparent activity of this class of agents is restricted to patients with renal carcinomas in which pVHL function is compromised. However, VEGF over-production in kidney cancer is more common than *VHL* inactivation, suggesting that some tumors achieve increased VEGF synthesis through mutations of genes other than *VHL* (Takahashi et al. 1994; Nicol et al. 1997).

Platelet-Derived Growth Factor

Platelet-derived growth factor B (PDGF-B), the ligand for PDGFR, is a hypoxia and HIF-responsive growth factor that is critical for the survival of pericytes which support the maintenance of endothelial cells (Kourembanas et al. 1990; Iliopoulos et al. 1996; Bergers and Benjamin 2003). While immature blood vessels, composed of only endothelial cells, appear to be exquisitely sensitive to VEGF withdrawl, mature blood vessels, because of the presence of surrounding pericytes, are much less sensitive to the same conditions (Bergers et al. 2003). Laboratory experiments suggest that involution of established blood vessels requires dual inhibition of VEGF and PDGF (Benjamin et al. 1999; Bergers et al. 2003). In this regard, KDR and PDGFR are phylogenetically and structurally related. As a result, many of the KDR inhibitors described above also inhibit the PDGF receptor.

Imatinib mesylate (Gleevec, Novartis), which is used to treat chronic myeloid leukemia (CML) and gastrointestinal stromal tumors (GIST) because of its activity against the c-Abl and c-Kit kinases, respectively, is also a potent inhibitor of PDGFR. As a single agent, imantinib did not induce any partial or complete responses in a small phase II trial of patients with metastatic renal cell carcinoma (Vuky et al. 2006). It is currently being tested for treatment of renal cell carcinoma in combination with VEGF/KDR inhibitors. While c-Kit overexpression by immuno-histochemistry has been reported in the sarcomatoid, oncocytoma, and chromophobe subtypes of renal cell carcinoma, only a small fraction (less than 5%) of clear cell renal cell carcinoma appear to overproduce c-Kit or have c-Kit mutations, making this an unlikely therapeutic target (Castillo et al. 2004; Kruger et al. 2005; Zigeuner et al. 2005; Sengupta et al. 2006; Vuky et al. 2006).

Transforming Growth Factor-α and Epidermal Growth Factor Receptor

Transforming growth factor alpha (TGFα) is also a HIF-responsive growth factor that mediates an autocrine loop in renal carcinoma cells by activating epidermal growth factor receptors (EGFRs) expressed on their surface (Knebelmann et al. 1998; de Paulsen et al. 2001; Gunaratnam et al. 2003). Normal renal epithelial cells are very sensitive to the mitogenic effects of TGFα (de Paulsen et al. 2001). The small molecule inhibitors, erlotinib (Tarceva, Genetech) and gefitinib (Iressa, AstraZeneca), and monoclonal antibodies, cetuximab (Erubitux, Imclone), targeting the EGFR are now FDA approved for other indications.

Despite solid rationale and suggestive preclinical studies, EGFR inhibitors tested as single agents in metastatic renal cell carcinoma have thus far displayed limited activity (Motzer et al. 2003; Rowinsky et al. 2004). As well, a recent report from a randomized phase II trial of bevacizumab with or without erlotinib in patients with metatatic renal cell carcinoma found no benefit to the combination over bevacizumab alone (Bukowski et al. 2007). In non-small cell lung cancer, response to EGFR inhibition is linked to the presence of activating mutations within the kinase domain of EGFR (Lynch et al. 2004; Paez et al. 2004). Whether this paradigm holds true for renal cell carcinoma is presently unknown. To date, there is little information about whether EGFR mutations exist in renal cell carcinoma with the exception of a small series of Japanese patients in whom no EGFR muta-tions were detected (Sakaeda et al. 2005).

c-MET (Mesenchymal Epithelial Transition Factor)

A subset of hereditary papillary renal carcinomas are secondary to inherited mutations in the receptor tyrosine kinase mesenchymal epithelial transition factor (c-Met)

(Linehan et al. 2007). There is evidence that c-MET, and its ligand, hepatocyte growth factor (HGF), might play a role in clear cell renal carcinomagenesis as well. Both HGF and c-MET can be activated by hypoxia or HIF and have been shown to be overexpressed in clear cell renal cell carcinoma (Pisters et al. 1997; Pennacchietti et al. 2003; Yamauchi et al. 2004; Hayashi et al. 2005; Nakaigawa et al. 2006). However, mutations of c-MET have not been reported in the clear cell subtype (Schmidt et al. 1997). Small molecules that inhibit the kinase activity of c-MET have been developed and are in clinical trials for papillary renal cancer.

Erythropoietin (EPO)

Renal carcinomas are one of several tumors occasionally associated with the para-neoplastic syndrome of polycythemia (Kim and Kaelin 2004). EPO is a HIF target, and while approximately a third of renal cell carcinomas express Epo mRNA, few pVHL-defective renal carcinomas overproduce and secrete EPO (Wiesener et al. 2007). It has also been suggested that renal cell carcinomas may express the erythropoietin receptor (EpoR) (Westenfelder and Baranowski 2000; Lee et al. 2005; Gong et al. 2006). If this observation is corroborated, it would suggest that EPO acts as an autocrine growth factor and would warrant trials with EpoR antagonists.

Conclusions

The best characterized function of the pVHL tumor suppressor protein is its ability to downregulate HIFα subunits, an attribute necessary for its tumor suppressor function. The *VHL* gene is mutated or silenced by hypermethylation in the majority of clear cell renal carcinomas. Inactivation of pVHL and its ubiquitin ligase activity result in the stabilization of HIFα subunits and the transactivation of HIF target genes. Therapeutic approaches that restore pVHL action or inhibit HIF transcriptional activity remain unrealized and will likely remain a therapeutic challenge. Several pVHL independent pathways that alter HIFα protein stability have recently been identified and the availability of inhibitors in clinical trials make these pathways attractive for further investigation.

Several HIF target genes encode growth factors that activate receptor tyrosine kinases (RTKs). In turn, many of these RTKs are now amenable to therapeutic blockade by tyrosine kinase inhibitors (TKIs) and monoclonal antibodies. Outcomes of clinical trials using combinations of these agents have been mixed and may reflect a true lack of biological effectiveness or an inability to fully inhibit the targeted pathways. The combined inhibition of collateral pathways (horizontal combinations) by agents that inhibit different protumorigenic HIF targets (such as VEGF and c-MET) is a rational extension of the success of these single agents. An alternative approach would be to combine agents that indirectly inhibit HIF with agents which inhibit HIF-responsive growth factors and their receptors (vertical combinations). However, targeted therapy combinations need to be evaluated systematically, as combining agents may have additive toxic effects without additional benefit.

References

Akiyama, H., T. Tanaka, et al. (2004). "Inhibition of ocular angiogenesis by an adenovirus carrying the human von Hippel-Lindau tumor-suppressor gene in vivo". Invest Ophthalmol Vis Sci 45(5): 1289–96.

Atkins, M.B., M. Hidalgo, et al. (2004). "Randomized phase II study of multiple dose levels of CCI-779, a novel mammalian target of rapamycin kinase inhibitor, in patients with advanced refractory renal cell carcinoma". J Clin Oncol 22(5): 909–18.

Bai, C., P. Sen, et al. (1996). "SKP1 connects cell cycle regulators to the ubiquitin proteolysis machinery through a novel motif, the F-box." Cell 86(2): 263–74.

Benjamin, L.E., D. Golijanin, et al. (1999). "Selective ablation of immature blood vessels in established human tumors follows vascular endothelial growth factor withdrawal". J Clin Invest 103(2): 159–65.

Bergers, G. and L.E. Benjamin (2003). "Tumorigenesis and the angiogenic switch". Nat Rev Cancer 3(6): 401–10.

Bergers, G., S. Song, et al. (2003). "Benefits of targeting both pericytes and endothelial cells in the tumor vasculature with kinase inhibitors." J Clin Invest 111(9): 1287–95.

Blankenship, C., J.G. Naglich, et al. (1999). "Alternate choice of initiation codon produces a biologically active product of the von Hippel Lindau gene with tumor suppressor activity." Oncogene 18(8): 1529–35.

Bolden, J.E., M.J. Peart, et al. (2006). "Anticancer activities of histone deacetylase inhibitors." Nat Rev Drug Discov 5(9): 769–84.

Brugarolas, J.B., F. Vazquez, et al. (2003). "TSC2 regulates VEGF through mTOR-dependent and -independent pathways." Cancer Cell 4(2): 147–58.

Bruick, R.K.and S.L. McKnight (2001). "A conserved family of prolyl-4-hydroxylases that modify HIF." Science 294(5545): 1337–40.

Bukowski, R.M., F.F. Kabbinavar, et al. (2007). "Randomized phase II study of erlotinib combined with bevacizumab compared with bevacizumab alone in metastatic renal cell cancer." J Clin Oncol 25(29): 4536–41.

Caldwell, M.C., C. Hough, et al. (2002). "Serial analysis of gene expression in renal carcinoma cells reveals VHL-dependent sensitivity to TNFalpha cytotoxicity." Oncogene 21(6): 929–36.

Castillo, M., A. Petit, et al. (2004). "C-kit expression in sarcomatoid renal cell carcinoma: potential therapy with imatinib." J Urol 171(6 Pt 1): 2176–80.

Cockman, M.E., N. Masson, et al. (2000). "Hypoxia inducible factor-alpha binding and ubiquitylation by the von Hippel–Lindau tumor suppressor protein." J Biol Chem 275(33): 25733–41.

Costa, L.J.and H.A. Drabkin (2007). "Renal cell carcinoma: new developments in molecular biology and potential for targeted therapies." Oncologist 12(12): 1404–15.

de Paulsen, N., A. Brychzy, et al. (2001). "Role of transforming growth factor-alpha in von Hippel-Lindau (VHL) (–/–) clear cell renal carcinoma cell proliferation: a possible mechanism coupling VHL tumor suppressor inactivation and tumorigenesis." Proc Natl Acad Sci U S A 98(4): 1387–92.

Dham, A.and A.Z. Dudek (2007). "Sequential therapy with sorafenib and sunitinib in renal cell carcinoma." J Clin Oncol, ASCO Annual Meeting Proceedings Part I **25**(No. 18S (June 20 Supplement)): 5106.

Duan, D.R., A. Pause, et al. (1995). "Inhibition of transcription elongation by the VHL tumor suppressor protein." Science **269**(5229): 1402–6.

Epstein, A.C., J.M. Gleadle, et al. (2001). "C. elegans EGL-9 and mammalian homologs define a family of dioxygenases that regulate HIF by prolyl hydroxylation." Cell **107**(1): 43–54.

Escudier, B., T. Eisen, et al. (2007). "Sorafenib in advanced clear-cell renal-cell carcinoma." N Engl J Med **356**(2): 125–34.

Escudier, B., A. Pluzanska, et al. (2007). "Bevacizumab plus interferon alfa-2a for treatment of metastatic renal cell carcinoma: a randomised, double-blind phase III trial." Lancet **370**(9605): 2103–11.

Fox, S.B., H. Turley, et al. (2004). "Phosphorylated KDR is expressed in the neoplastic and stromal elements of human renal tumours and shuttles from cell membrane to nucleus." J Pathol **202**(3): 313–20.

Galban, S., J. Fan, et al. (2003). "von Hippel–Lindau protein-mediated repression of tumor necrosis factor alpha translation revealed through use of cDNA arrays." Mol Cell Biol **23**(7): 2316–28.

George, P., P. Bali, et al. (2004). "Cotreatment with 17-allylamino-demethoxygeldanamycin and FLT-3 kinase inhibitor PKC412 is highly effective against human acute myelogenous leukemia cells with mutant FLT-3." Cancer Res **64**(10): 3645–52.

Gong, K., N. Zhang, et al. (2006). "Coexpression of erythopoietin and erythopoietin receptor in sporadic clear cell renal cell carcinoma." Cancer Biol Ther **5**(6): 582–5.

Gradin, K., J. McGuire, et al. (1996). "Functional interference between hypoxia and dioxin signal transduction pathways: competition for recruitment of the Arnt transcription factor." Mol Cell Biol **16**(10): 5221–31.

Gunaratnam, L., M. Morley, et al. (2003). "Hypoxia inducible factor activates the transforming growth factor-alpha/epidermal growth factor receptor growth stimulatory pathway in VHL (–/–) renal cell carcinoma cells." J Biol Chem **278**(45): 44966–74.

Hacein-Bey-Abina, S., C. Von Kalle, et al. (2003). "LMO2-associated clonal T cell proliferation in two patients after gene therapy for SCID-X1." Science **302**(5644): 415–9.

Hayashi, M., M. Sakata, et al. (2005). "Up-regulation of c-met protooncogene product expression through hypoxia-inducible factor-1alpha is involved in trophoblast invasion under low-oxygen tension." Endocrinology **146**(11): 4682–9.

Hergovich, A., J. Lisztwan, et al. (2003). "Regulation of microtubule stability by the von Hippel–Lindau tumour suppressor protein pVHL." Nat Cell Biol **5**(1): 64–70.

Hergovich, A., J. Lisztwan, et al. (2006). "Priming-dependent phosphorylation and regulation of the tumor suppressor pVHL by glycogen synthase kinase 3." Mol Cell Biol **26**(15): 5784–96.

Hood, J.D., M. Bednarski, et al. (2002). "Tumor regression by targeted gene delivery to the neovasculature." Science **296**(5577): 2404–7.

Huang, L.E., J. Gu, et al. (1998). "Regulation of hypoxia-inducible factor 1alpha is mediated by an O2-dependent degradation domain via the ubiquitin-proteasome pathway." Proc Natl Acad Sci U S A **95**(14): 7987–92.

Hudes, G., M. Carducci, et al. (2007). "Temsirolimus, interferon alfa, or both for advanced renal-cell carcinoma." N Engl J Med **356**(22): 2271–81.

Hudson, C.C., M. Liu, et al. (2002). "Regulation of hypoxia-inducible factor 1alpha expression and function by the mammalian target of rapamycin." Mol Cell Biol **22**(20): 7004–14.

Hutson, T.E., I.D. Davis, et al. (2007). "Pazopanib (GW786034) is active in metastatic renal cell carcinoma (RCC): Interim results of a phase II randomized discontinuation trial (RDT)." J Clin Oncol, ASCO Annual Meeting Proceedings **25** (June 20 Supplement): 5031.

Iliopoulos, O., A. Kibel, et al. (1995). "Tumour suppression by the human von Hippel–Lindau gene product." Nat Med **1**(8): 822–6.

Iliopoulos, O., A.P. Levy, et al. (1996). "Negative regulation of hypoxia-inducible genes by the von Hippel–Lindau protein." Proc Natl Acad Sci U S A **93**(20): 10595–9.

Iliopoulos, O., M. Ohh, et al. (1998). "pVHL19 is a biologically active product of the von Hippel–Lindau gene arising from internal translation initiation." Proc Natl Acad Sci U S A **95**(20): 11661–6.

Isaacs, J.S., Y.J. Jung, et al. (2002). "Hsp90 regulates a von Hippel Lindau-independent hypoxia-inducible factor-1 alpha-degradative pathway." J Biol Chem **277**(33): 29936–44.

Isaacs, J.S., W. Xu, et al. (2003). "Heat shock protein 90 as a molecular target for cancer therapeutics." Cancer Cell **3**(3): 213–7.

Ivan, M., T. Haberberger, et al. (2002). "Biochemical purification and pharmacological inhibition of a mammalian prolyl hydroxylase acting on hypoxia-inducible factor." Proc Natl Acad Sci U S A **99**(21): 13459–64.

Ivan, M., K. Kondo, et al. (2001). "HIFalpha targeted for VHL-mediated destruction by proline hydroxylation: implications for O2 sensing." Science **292**(5516): 464–8.

Jaakkola, P., D.R. Mole, et al. (2001). "Targeting of HIF-alpha to the von Hippel–Lindau ubiquitylation complex by O2-regulated prolyl hydroxylation." Science **292**(5516): 468–72.

Jeong, J.W., M.K. Bae, et al. (2002). "Regulation and destabilization of HIF-1alpha by ARD1-mediated acetylation." Cell **111**(5): 709–20.

Jiang, Y., W. Zhang, et al. (2003). "Gene expression profiling in a renal cell carcinoma cell line: dissecting VHL and hypoxia-dependent pathways." Mol Cancer Res **1**(6): 453–62.

Kaelin, W.G. (2005). "von Hippel–Lindau-associated malignancies: Mechanisms and therapeutic opportunities." Drug Discov Today **2**(2): 225–31.

Kaelin, W.G., Jr. (2002). "Molecular basis of the VHL hereditary cancer syndrome." Nat Rev Cancer **2**(9): 673–82.

Kamura, T., M.N. Conrad, et al. (1999). "The Rbx1 subunit of SCF and VHL E3 ubiquitin ligase activates Rub1 modification of cullins Cdc53 and Cul2." Genes Dev **13**(22): 2928–33.

Kamura, T., S. Sato, et al. (2000). "Activation of HIF1alpha ubiquitination by a reconstituted von Hippel–Lindau (VHL) tumor suppressor complex." Proc Natl Acad Sci U S A **97**(19): 10430–5.

Kibel, A., O. Iliopoulos, et al. (1995). "Binding of the von Hippel-Lindau tumor suppressor protein to Elongin B and C." Science **269**(5229): 1444–6.

Kim, M.S., H.J. Kwon, et al. (2001). "Histone deacetylases induce angiogenesis by negative regulation of tumor suppressor genes." Nat Med **7**(4): 437–43.

Kim, W.Y. and W.G. Kaelin (2004). "Role of VHL gene mutation in human cancer." J Clin Oncol **22**(24): 4991–5004.

Kim, W.Y., M. Safran, et al. (2006). "Failure to prolyl hydroxylate HIFalpha phenocopies VHL inactivation in vivo." Embo J **25**(19): 4650–62.

Kishida, T., T.M. Stackhouse, et al. (1995). "Cellular proteins that bind the von Hippel-Lindau disease gene product: mapping of binding domains and the effect of missense mutations." Cancer Res **55**(20): 4544–8.

Knebelmann, B., S. Ananth, et al. (1998). "Transforming growth factor alpha is a target for the von Hippel–Lindau tumor suppressor." Cancer Res **58**(2): 226–31.

Knudson, A.G., Jr. (1971). "Mutation and cancer: statistical study of retinoblastoma." Proc Natl Acad Sci U S A **68**(4): 820–3.

Kondo, K., W.Y. Kim, et al. (2003). "Inhibition of HIF2alpha is sufficient to suppress pVHL-defective tumor growth." PLoS Biol **1**(3): E83.

Kondo, K., J. Klco, et al. (2002). "Inhibition of HIF is necessary for tumor suppression by the von Hippel–Lindau protein." Cancer Cell **1**(3): 237–46.

Kong, X., Z. Lin, et al. (2006). "Histone deacetylase inhibitors induce VHL and ubiquitin-independent proteasomal degradation of hypoxia-inducible factor 1{alpha}." Mol Cell Biol **26**(6): 2019–28.

Kourembanas, S., R.L. Hannan, et al. (1990). "Oxygen tension regulates the expression of the platelet-derived growth factor-B chain gene in human endothelial cells." J Clin Invest **86**(2): 670–4.

Kruger, S., K. Sotlar, et al. (2005). "Expression of KIT (CD117) in renal cell carcinoma and renal oncocytoma." Oncology 68(2–3): 269–75.

Latif, F., K. Tory, et al. (1993). "Identification of the von Hippel-Lindau disease tumor suppressor gene." Science 260(5112): 1317–20.

Lee, S., D.Y. Chen, et al. (1996). "Nuclear/cytoplasmic localization of the von Hippel-Lindau tumor suppressor gene product is determined by cell density." Proc Natl Acad Sci U S A 93(5): 1770–5.

Lee, S., M. Neumann, et al. (1999). "Transcription-dependent nuclear-cytoplasmic trafficking is required for the function of the von Hippel–Lindau tumor suppressor protein." Mol Cell Biol 19(2): 1486–97.

Lee, Y.S., A.O. Vortmeyer, et al. (2005). "Coexpression of erythropoietin and erythropoietin receptor in von Hippel–Lindau disease-associated renal cysts and renal cell carcinoma." Clin Cancer Res 11(3): 1059–64.

Linehan, W.M., P.A. Pinto, et al. (2007). "Identification of the genes for kidney cancer: opportunity for disease-specific targeted therapeutics." Clin Cancer Res 13(2 Pt 2): 671s–679s.

Liu, Y.V., J.H. Baek, et al. (2007). "RACK1 competes with HSP90 for binding to HIF-1alpha and is required for O(2)-independent and HSP90 inhibitor-induced degradation of HIF-1alpha." Mol Cell 25(2): 207–17.

Lolkema, M.P., M.L. Gervais, et al. (2005). "Tumor suppression by the von Hippel-Lindau protein requires phosphorylation of the acidic domain." J Biol Chem 280(23): 22205–11.

Lonergan, K.M., O. Iliopoulos, et al. (1998). "Regulation of hypoxia-inducible mRNAs by the von Hippel–Lindau tumor suppressor protein requires binding to complexes containing elongins B/C and Cul2." Mol Cell Biol 18(2): 732–41.

Lynch, T.J., D.W. Bell, et al. (2004). "Activating mutations in the epidermal growth factor receptor underlying responsiveness of non-small-cell lung cancer to gefitinib." N Engl J Med 350(21): 2129–39.

Mabjeesh, N.J., D. Escuin, et al. (2003). "2ME2 inhibits tumor growth and angiogenesis by disrupting microtubules and dysregulating HIF." Cancer Cell 3(4): 363–75.

Mabjeesh, N.J., D.E. Post, et al. (2002). "Geldanamycin induces degradation of hypoxia-inducible factor 1alpha protein via the proteosome pathway in prostate cancer cells." Cancer Res 62(9): 2478–82.

Mandriota, S.J., K.J. Turner, et al. (2002). "HIF activation identifies early lesions in VHL kidneys: evidence for site-specific tumor suppressor function in the nephron." Cancer Cell 1(5): 459–68.

Maranchie, J.K., J.R. Vasselli, et al. (2002). "The contribution of VHL substrate binding and HIF1-alpha to the phenotype of VHL loss in renal cell carcinoma." Cancer Cell 1(3): 247–55.

Masson, N., C. Willam, et al. (2001). "Independent function of two destruction domains in hypoxia-inducible factor-alpha chains activated by prolyl hydroxylation." Embo J 20(18): 5197–206.

Maxwell, P.H., M.S. Wiesener, et al. (1999). "The tumour suppressor protein VHL targets hypoxia-inducible factors for oxygen-dependent proteolysis." Nature 399(6733): 271–5.

Melillo, G. (2007). "Hypoxia-inducible factor 1 inhibitors." Methods Enzymol 435: 385–402.

Mendez, R., M.G. Myers, Jr., et al. (1996). "Stimulation of protein synthesis, eukaryotic translation initiation factor 4E phosphorylation, and PHAS-I phosphorylation by insulin requires insulin receptor substrate 1 and phosphatidylinositol 3-kinase." Mol Cell Biol 16(6): 2857–64.

Meyuhas, O. (2000). "Synthesis of the translational apparatus is regulated at the translational level." Eur J Biochem 267(21): 6321–30.

Motzer, R.J., R. Amato, et al. (2003). "Phase II trial of antiepidermal growth factor receptor antibody C225 in patients with advanced renal cell carcinoma." Invest New Drugs 21(1): 99–101.

Motzer, R.J., T.E. Hutson, et al. (2007). "Sunitinib versus interferon alfa in metastatic renal-cell carcinoma." N Engl J Med 356(2): 115–24.

Nakaigawa, N., M. Yao, et al. (2006). "Inactivation of von Hippel–Lindau gene induces constitutive phosphorylation of MET protein in clear cell renal carcinoma." Cancer Res 66(7): 3699–705.

Nicol, D., S.I. Hii, et al. (1997). "Vascular endothelial growth factor expression is increased in renal cell carcinoma." J Urol 157(4): 1482–6.

Ohh, M., C.W. Park, et al. (2000). "Ubiquitination of hypoxia-inducible factor requires direct binding to the beta-domain of the von Hippel–Lindau protein." Nat Cell Biol 2(7): 423–7.

Paez, J.G., P.A. Janne, et al. (2004). "EGFR mutations in lung cancer: correlation with clinical response to gefitinib therapy." Science 304(5676): 1497–500.

Pause, A., S. Lee, et al. (1997). "The von Hippel–Lindau tumor-suppressor gene product forms a stable complex with human CUL-2, a member of the Cdc53 family of proteins." Proc Natl Acad Sci U S A 94(6): 2156–61.

Pennacchietti, S., P. Michieli, et al. (2003). "Hypoxia promotes invasive growth by transcriptional activation of the met protooncogene." Cancer Cell 3(4): 347–61.

Pisters, L.L., A.K. el-Naggar, et al. (1997). "C-met proto-oncogene expression in benign and malignant human renal tissues." J Urol 158(3 Pt 1): 724–8.

Qian, D.Z., S.K. Kachhap, et al. (2006). "Class II histone deacetylases are associated with VHL-independent regulation of hypoxia-inducible factor 1 alpha." Cancer Res 66(17): 8814–21.

Rahmani, M., C. Yu, et al. (2003). "Coadministration of the heat shock protein 90 antagonist 17-allylamino-17-demethoxygeldanamycin with suberoylanilide hydroxamic acid or sodium butyrate synergistically induces apoptosis in human leukemia cells." Cancer Res 63(23): 8420–7.

Rankin, E.B., D.F. Higgins, et al. (2005). "Inactivation of the arylhydrocarbon receptor nuclear translocator (Arnt) suppresses von Hippel–Lindau disease-associated vascular tumors in mice." Mol Cell Biol 25(8): 3163–72.

Rankin, E.B., J.E. Tomaszewski, et al. (2006). "Renal cyst development in mice with conditional inactivation of the von Hippel–Lindau tumor suppressor." Cancer Res 66(5): 2576–83.

Rapisarda, A., B. Uranchimeg, et al. (2004). "Topoisomerase I-mediated inhibition of hypoxia-inducible factor 1: mechanism and therapeutic implications." Cancer Res 64(4): 1475–82.

Rixe, O., R.M. Bukowski, et al. (2007). "Axitinib treatment in patients with cytokine-refractory metastatic renal-cell cancer: a phase II study." Lancet Oncol 8(11): 975–84.

Ronnen, E.A., G.V. Kondagunta, et al. (2006). "A phase II trial of 17-(Allylamino)-17-demethoxygeldanamycin in patients with papillary and clear cell renal cell carcinoma." Invest New Drugs 24(6): 543–6.

Rosenwald, I.B., R. Kaspar, et al. (1995). "Eukaryotic translation initiation factor 4E regulates expression of cyclin D1 at transcriptional and post-transcriptional levels." J Biol Chem 270(36): 21176–80.

Rowinsky, E.K., G.H. Schwartz, et al. (2004). "Safety, pharmacokinetics, and activity of ABX-EGF, a fully human anti-epidermal growth factor receptor monoclonal antibody in patients with metastatic renal cell cancer." J Clin Oncol 22(15): 3003–15.

Sakaeda, T., N. Okamura, et al. (2005). "EGFR mRNA is upregulated, but somatic mutations of the gene are hardly found in renal cell carcinoma in Japanese patients." Pharm Res 22(10): 1757–61.

Schmidt, L., F.M. Duh, et al. (1997). "Germline and somatic mutations in the tyrosine kinase domain of the MET proto-oncogene in papillary renal carcinomas." Nat Genet 16(1): 68–73.

Schoenfeld, A., E.J. Davidowitz, et al. (1998). "A second major native von Hippel–Lindau gene product, initiated from an internal translation start site, functions as a tumor suppressor." Proc Natl Acad Sci U S A 95(15): 8817–22.

Seizinger, B.R., G.A. Rouleau, et al. (1988). "Von Hippel–Lindau disease maps to the region of chromosome 3 associated with renal cell carcinoma." Nature 332(6161): 268–9.

Semenza, G.L. (2003). "Targeting HIF-1 for cancer therapy." Nat Rev Cancer 3(10): 721–32.

Sengupta, S., J.C. Cheville, et al. (2006). "Rare expression of KIT and absence of KIT mutations in high grade renal cell carcinoma." J Urol 175(1): 53–6.

Shaw, R.J. and L.C. Cantley (2006). "Ras, PI(3)K and mTOR signalling controls tumour cell growth." Nature **441**(7092): 424–30.

Solit, D.B., S.P. Ivy, et al. (2007). "Phase I trial of 17-allylamino-17-demethoxygeldanamycin in patients with advanced cancer." Clin Cancer Res **13**(6): 1775–82.

Stebbins, C.E., W.G. Kaelin, Jr., et al. (1999). "Structure of the VHL-ElonginC-ElonginB complex: implications for VHL tumor suppressor function." Science **284**(5413): 455–61.

Takahashi, A., H. Sasaki, et al. (1994). "Markedly increased amounts of messenger RNAs for vascular endothelial growth factor and placenta growth factor in renal cell carcinoma associated with angiogenesis." Cancer Res **54**(15): 4233–7.

Tanimoto, K., Y. Makino, et al. (2000). "Mechanism of regulation of the hypoxia-inducible factor-1 alpha by the von Hippel–Lindau tumor suppressor protein." Embo J **19**(16): 4298–309.

Thomas, G.V. (2006). "mTOR and cancer: Reason for dancing at the crossroads?" Curr Opin Genet Dev **16**(1): 78–84.

Thomas, G.V., C. Tran, et al. (2006). "Hypoxia-inducible factor determines sensitivity to inhibitors of mTOR in kidney cancer." Nat Med **12**(1): 122–7.

Treins, C., S. Giorgetti-Peraldi, et al. (2002). "Insulin stimulates hypoxia-inducible factor 1 through a phosphatidylinositol 3-kinase/target of rapamycin-dependent signaling pathway." J Biol Chem **277**(31): 27975–81.

Vaishampayan, U., E. Sausville, et al. (2007). "Phase I trial of intravenous 17-allylaminogeldanamycin (A) and oral sorafenib (B) in pretreated advanced malignancy: Plasma Hsp90α induction correlates with clinical benefit." Am Soc Clin Oncol **25**(185): 3531.

Vuky, J., C. Isacson, et al. (2006). "Phase II trial of imatinib (Gleevec) in patients with metastatic renal cell carcinoma." Invest New Drugs **24**(1): 85–8.

Westenfelder, C. and R.L. Baranowski (2000). "Erythropoietin stimulates proliferation of human renal carcinoma cells." Kidney Int **58**(2): 647–57.

Wiesener, M.S., P. Munchenhagen, et al. (2007). "Erythropoietin gene expression in renal carcinoma is considerably more frequent than paraneoplastic polycythemia." Int J Cancer **121**(11): 2434–42.

Yamauchi, M., H. Kataoka, et al. (2004). "Hepatocyte growth factor activator inhibitor types 1 and 2 are expressed by tubular epithelium in kidney and down-regulated in renal cell carcinoma." J Urol **171**(2 Pt 1): 890–6.

Yang, J.C., L. Haworth, et al. (2003). "A randomized trial of bevacizumab, an anti-vascular endothelial growth factor antibody, for metastatic renal cancer." N Engl J Med **349**(5): 427–34.

Yu, F., S.B. White, et al. (2001). "HIF-1alpha binding to VHL is regulated by stimulus-sensitive proline hydroxylation." Proc Natl Acad Sci U S A **98**(17): 9630–5.

Zigeuner, R., M. Ratschek, et al. (2005). "Kit (CD117) immunoreactivity is rare in renal cell and upper urinary tract transitional cell carcinomas." BJU Int **95**(3): 315–8.

Zimmer, M., D. Doucette, et al. (2004). "Inhibition of hypoxia-inducible factor is sufficient for growth suppression of VHL–/– tumors." Mol Cancer Res **2**(2): 89–95.

Interrelationship of the Fanconi Anemia/BRCA Pathway

Patricia McChesney and Gary M. Kupfer

Abstract Breast cancer is the leading cause of cancer in women, and the second leading cause of cancer-related death in women. Approximately 10% of women diagnosed with breast cancer have some familial history of the disease. Prevalence of breast cancer in some families led to the discoveries of BRCA1 and BRCA2, the well-established breast cancer susceptibility genes. Loss or mutation of BRCA1 or BRCA2 results in widespread genomic instability that is strikingly similar to the genomic instability found in the inherited disease Fanconi anemia (FA).

The similarity has led researchers to discover that BRCA1 and BRCA2 are integral players in a DNA repair pathway that involves the 13 members in the FA family of proteins with BRCA2 being itself an FA gene.

This review will provide insight into how the FA/BRCA pathway functions, and will discuss therapeutic opportunities for the future.

Keywords BRCA1/2 • Fanconi anemia • DNA damage • DNA repair • Genomic instability • Monoubiquitylation

Introduction

Breast cancer is the leading type of cancer in women and the second leading cause of cancer-related deaths in women. Approximately ten percent of women diagnosed with breast cancer have some familial history of the disease (Boulton 2006). Prevalence of breast cancer in some families led to the discoveries of BRCA1 and BRCA2 (Hall et al. 1990; Miki et al. 1994; Wooster et al. 1994; Wooster et al. 1995), the well-established breast cancer susceptibility genes. Loss or mutation of BRCA1 or BRCA2 results in widespread genomic instability which is strikingly similar to the genomic instability found in the inherited disease

P. McChesney and G.M. Kupfer (✉)

Department of Pediatrics, Section of Pediatric Hematology-Oncology,
Yale University School of Medicine, New Haven, CT, USA

K. Sakamoto and E. Rubin (eds.), *Modulation of Protein Stability in Cancer Therapy*,
DOI: 10.1007/978-0-387-69147-3_5, © Springer Science+Business Media, LLC 2009

Fanconi anemia (FA). The similarity has led researchers to discover that BRCA1 and BRCA2 are integral players in a DNA repair pathway that involves the thirteen members in the FA family of proteins with BRCA2 being itself an FA gene. This review provides insight into how the FA/BRCA pathway functions and discusses therapeutic opportunities for the future.

Familial Breast Cancer

Familial breast cancer is distinguished from sporadic breast cancer by the younger age at diagnosis (less than 50-years old), frequent bilateral disease, and frequent occurrence of disease among men (Hall et al. 1990). Linkage analysis among affected families led to the identification of BRCA1 as a tumor suppressor gene (Hall et al. 1990). Subsequent sequence analysis has shown that truncating mutations account for many cases and over 300 missense sequence mutations have been identified (Boulton 2006). However, BRCA1-related cancers only affect women and could not explain familial breast cancer where men were also afflicted. Thus, the identification of BRCA2 was made (Wooster et al. 1994; Wooster et al. 1995). In the BRCA2 study, fifteen breast cancer families could not be identified as BRCA1 mutant; had 146 cases of breast cancer, with 71% being in females under the age 50 and 5.4% being in males (Wooster et al. 1994). Consistent with other tumor suppressor genes, cancer generally arose because of loss of heterozygosity (LOH) and loss of the wild-type allele (Narod 2002; Narod and Foulkes 2004; Turner et al. 2004).

The statistical occurrence of familial breast cancer is approximately 10% of all diagnosed cases, with considerably higher inherited mutations in patients who are less than 50-years old. Of the families with inherited breast cancer, BRCA1 mutations account for more than half, while BRCA2 mutations account for less than one third (Rebbeck et al. 1996; Ford et al. 1998). Interestingly, of the families with reported cases of male breast cancer, 76% were BRCA2 mutant (Ford et al. 1998). However, some cases of inherited breast cancer cannot be attributed to BRCA1 or BRCA2, and gene members of the FA family of proteins cause many of the exceptions.

Fanconi Anemia

Fanconi anemia (FA) is a rare genetic disease, characterized by chromosomal instability, bone marrow failure, congenital defects, solid tumors, and leukemia (Fanconi 1967; Alter and Young 1993; Auerbach et al. 1997; Alter 2003). The well-documented association of FA with DNA damage hypersensitivity led investigators to classify FA as a DNA repair defect disorder (Alter 2003; Collins and Kupfer 2005), specifically as a condition with deficient repair of DNA interstrand crosslinks (ICL).

Interstrand DNA Crosslinks

Interstrand DNA crosslinks (ICLs) covalently tether both strands of the DNA helix together. This prevents strand separation, a requirement for cellular processes such as DNA replication and transcription. Consequently, ICLs are extremely cytotoxic, particularly for proliferating cells. Hence, crosslinking chemotherapeutic agents are among the most effective cytotoxic drugs used in cancer chemotherapy. *In vivo*, replication of DNA containing ICLs induces double-strand breaks (DSBs) (De Silva et al. 2000; Niedernhofer et al. 2004). Growing evidence suggests that ICL repair occurs in a coordinate fashion with DNA replication and involves homologous recombination. ICLs also induce sister-chromatid exchange (SCE), suggesting that replication-dependent DSBs are often resolved through SCEs. SCE entails the reciprocal exchange of the DNA strands between identical sister chromatids and thus only occurs during or after DNA replication. Not surprisingly, SCE requires the homologous recombination machinery (Sonoda et al. 1999). There is a growing body of evidence suggesting that homologous recombination in the repair of DNA ICLs is regulated by the FA/BRCA pathway (Fig. 1).

Overview of the FA/BRCA Pathway

The FA/BRCA pathway is activated by DNA ICLs and begins with coordinated formation of FA protein complexes (Abu-Issa et al. 1999; Howlett et al. 2002). Thirteen FA genes have been identified; FANC-A, B, C, D1, D2, E, F, G, I, J, L, M,

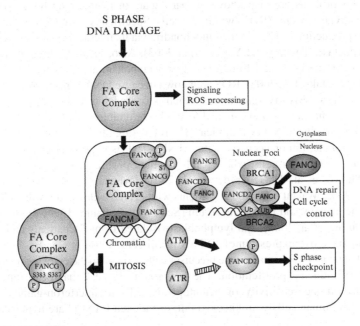

Fig. 1 The FA/BRCA pathway

and N (Strathdee et al. 1992; Whitney et al. 1995; Lo Ten Foe et al. 1996; de Winter et al. 1998, 2000; Hejna et al. 2000; Timmers et al. 2001; Meetei Winter et al. 2003; Meetei et al. 2005; Taniguchi and D'Andrea 2006). Biochemical studies of these proteins have shown that upon DNA damage a "core complex" forms in a stepwise manner. It begins in the cytoplasm with binding of FANC-A, B, C, F, and G. These proteins, once bound to each other, move into the nucleus (Thomashevski et al. 2004) and are joined by FANC-E, L, and M (Kupfer et al. 1997; Naf et al. 1998; Yamashita et al. 1998; Garcia-Higuera et al. 1999; Waisfisz et al. 1999; Christianson and Bagby 2000; de Winter et al. 2000; Hussain et al. 2003; Thomashevski et al. 2004). This complex is transferred to DNA via the DNA-binding activity of FANCM, where it participates in monoubiquitylation of FANCD2 and FANCI (Meetei et al. 2004; Wang et al. 2004; Smogorzewska et al. 2007). Interestingly, both ultraviolet radiation and ionizing radiation can also lead to monoubiquitylation of FANCD2, but only if applied during S-phase of the cell cycle (Houghtaling et al. 2003, 2005; Dunn et al. 2006). IR leads to monoubiquitylation of FANCD2, and mice with an FA gene knocked out have increased sensitivity to IR (Houghtaling et al. 2003; Houghtaling et al. 2005). Shortly after damage, Ub-FANCD2 is seen at presumed chromosome repair foci, along with BRCA1, RPA, PCNA, RAD51, BRCA2/FANCD1, and others (Andreassen et al. 2004; Wang et al. 2004; Kitao et al. 2006). Currently, BRCA2/FANCD1 is thought to act as a scaffold where the repair proteins and FANCN/PALB2 bind to fulfill their functions of DNA repair in a coordinated manner. BRCA2/FANCD1 binds the DNA repair proteins, including RAD51, while simultaneously binding both single- and double-stranded DNA (Taniguchi and D'Andrea 2006), suggesting that BRCA2/FANCD1 is a pivotal factor in DNA repair.

The FA proteins are also known to participate in maintaining homeostasis of reactive oxygen species (ROS) within the cells. Specifically, FANCC directly regulates the production of ROS in the mitochondria via its interaction with cytochrome P450 reductase (Pagano and Youssoufian 2003). Also, FANCG directly interacts with cytochrome P450 2E1 to regulate drug metabolism and thus reduce intracellular ROS (Futaki et al. 2002). ROS are known to cause DNA damage by converting guanosine into 8-hydroxy-guanosine, which is unreadable by DNA replication or transcription machinery, and can lead to single- or double-strand breaks.

The FA proteins have been implicated in regulating the expression of the hematologic cytokine, tumor necrosis factor (TNF). TNF is expressed and detected by hematologic precursor cells, where it inhibits their proliferation (Broxmeyer et al. 1986; Murase et al. 1987; Bryder et al. 2001; Dybedal et al. 2001; Chen and Goeddel 2002; Wajant et al. 2003). Patients with FA have elevated levels of TNF in their bone marrow and peripheral serum (Schultz and Shahidi 1993; Rosselli et al. 1994; Dufour et al. 2003), and lymphoblasts isolated from FA patients and grown in culture have overexpression of TNF (Rosselli et al. 1994).

In addition to TNF, hematopoetic precursor cells respond to the cytokine interferon-gamma by reducing proliferation. The FA protein, FANCC has been repeatedly implicated in the sensitivity of hematopoetic cells to interferon-gamma. Both murine and human cells with knocked out, or mutated FANCC are hypersensitive to the growth inhibition effects of interferon-gamma (Whitney et al. 1996; Rathbun et al. 1997; Pang et al. 2001; Fagerlie and Bagby 2006). Interestingly, the sensitive response of human cells varied depending on the specific FANCC mutation involved (Pang et al. 2001).

Evidence is building that the FA/BRCA pathway may extend beyond DNA repair. In the next section of this chapter the multifaceted nature of each of the FA/BRCA proteins is reviewed.

The FA/BRCA Proteins

The thirteen subclasses of FA were originally established by complementation assays, in which cells from two patients would be fused together and assayed for DNA damage sensitivity. If the fused cells demonstrated reduced sensitivity, then the genotypes "complemented" each other and was thus evidence for two distinct complementation groups and presumably two different mutant genes (Levitus et al. 2004). All the 13 of these groups have accompanying genes cloned and these are described below in alphabetical order. Additional complementation groups may exist, suggesting more genes yet to be described.

FA subtype A (FA-A) accounts for approximately 60% of all FA patients, whose gene is designated FANCA (Kennedy and D'Andrea 2005). FANCA was identified by positional and expression cloning (1996; Foe et al. 1996) and like most of the FA proteins does not have recognizable functional motifs or domains. FANCA is a large protein at 163 kDa (Wang 2007) with two nuclear localization signals and a partial leucine zipper. FANCA is a member of the core complex, which is known to facilitate monoubiquitination of FANCD2. FANCA is also phosphorylated in a variety of locations (Yamashita et al. 1998; Thomashevski et al. 2004), each of which is required for its function in the FA pathway, but regulation of phosphorylation remains unknown.

FA-B accounts for less than 2% of FA cases reported till date (Kennedy and D'Andrea 2005) and is the only one to display X-linked inheritance (Meetei et al. 2004). FANCB is 95 kDa and also contains a nuclear localization sequence (Wang 2007). Interestingly, monoallelic carriers of mutated FANCB present with FA because of X chromosome inactivation that is entirely skewed toward the mutant allele (Meetei et al. 2004), suggesting a type of in vivo selection that has made FA generally a candidate disease for gene therapy. Like FANCA, it has no recognizable functional motifs or domains. FANCB is also known to participate in the FA core complex.

FA-C accounts for 15% of FA cases (Kennedy and D'Andrea 2005) and features the first of the identified FA genes, FANCC, a 63 kDa protein (Wang 2007). Sequencing of the mutant genes from patients has revealed that most mutations are either an intronic mutation that results in splice removal of exon 4 (IVS4 + 4A to T), or a truncation mutant in exon 1. The exon 4 mutation results in loss of FANCC expression, whereas the exon 1 mutation allows for downstream reinitiation and expression of a partially functional FANCC fragment, and thus a less severe patient phenotype (Yamashita et al. 1996). Like FANCA and FANCB, FANCC does not have recognizable functional motifs or domains and participates in the FA core complex.

FA subtype D accounts for a small fraction of FA cases. It was split into two subcategories when complementation data showed that two separate genes were responsible for this group (FANCD1 and FANCD2).

FANCD1 was identified as BRCA2 and was the first FA protein to be directly associated with familial breast cancer (Howlett et al. 2002). BRCA2 (Wooster et al 1995) mutations account for the majority of families where male breast cancer occurs in addition to female breast cancer (Boulton 2006). FANCD1/BRCA2 is 380 kDa and is conserved among vertebrates and worms (Wang 2007). Interestingly, heterozygous carriers of BRCA2/FA-D1 mutations are prone to breast and ovarian cancers, but not to malignancies normally associated with FA, such as squamous cell carcinoma of the head and neck, or AML. FANCD1/BRCA2 binds to DNA and mediates binding of other factors associated with DNA repair. It is not associated with the FA core complex, and is thought to function downstream of FANCD2 in the FA/BRCA pathway (Siddique et al. 2001) since FANCD2 is monoubiquitylated in this mutant group. BRCA2 gene encodes a protein which is structurally and functionally unrelated to BRCA1. BRCA2 protein is 3,329aa, with eight BRC repeats, and a multifunctional C-terminal domain with five subdomains (Taniguchi and D'Andrea 2006); one helical domain, three oligonucleotide/oligosaccharide biding folds (OB1, 2, and 3) that bind to ssDNA, and one tower domain (TD) that binds to dsDNA. BRCA2 seems to function primarily as a structural scaffold that brings multiple other proteins, along with DNA, into close proximity for interactions. It stabilizes stalled replication forks (likely via binding ssDNA and dsDNA simultaneously), and regulates homologous recombination repair through the control of RAD51 via its 8 BRC repeats (may aid in transfer of RAD51 to DNA).

FANCD2 is one of the only two FA proteins with homologues in lower eukaryotes such as worms, insects, and slime mold (Timmers et al. 2001). However, despite being well conserved, FANCD2 has no recognizable functional motifs or domains. FANCD2 is 155 kDa (Wang 2007). FANCD2 functions downstream of the FA core complex and its activation is dependent on the core complex. Upon DNA damage or S-phase of the cell cycle, FANCD2 is monoubiquitylated at lysine 561, causing a highly recognizable upward shift in the molecular weight of FANCD2 to 162 kDa (Wang 2007). This modification is necessary for the participation of FANCD2 in DNA repair complexes which is shown by immunofluorescence microscopy as discrete foci at the site of DNA damage or at the site of stalled replication forks. Proteins thought to be localized to repair foci also include, but are not limited to, BRCA1, Rad51, BLM, and histone gamma-H2AX. FANCD2 has been shown to immunoprecipitate with BRCA1 (Garcia-Higuera et al. 2001; Vandenberg et al. 2003), and histone gamma-H2AX (Bogliolo et al. 2007). FANCD2 is also phosphorylated at serine 222 by ATM kinase (Taniguchi et al. 2002), threonine 691, and serine 717; reactions that are regulated by ATM or ATR (Ho et al. 2006). Of these, only phosphorylation at serine 717 and threonine 691 are necessary for normal FA pathway function and ICL resistance.

FA-E is seen in less than 5% of FA patients. The gene mutated in this disorder is FANCE and as is true of most FA proteins; FANCE has no recognizable motifs or domains except for two nuclear localization signals (de Winter et al. 2000). FANCE is 60 kDa (Wang 2007). FANCE is the only FA protein known to coprecipitate with both FA core complex and FANCD2, albeit not simultaneously (Pace et al. 2002; Smogorzewska et al. 2007). As a result, FANCE is thought to be a molecular link between the core complex and FANCD2. Interestingly, while FANCE is required for monoubiquitylation of FANCD2, FANCE has two Chk1 kinase-dependent phosphorylation sites that are required to complement MMC sensitivity in FA-E cells but are not required for monoubiquitination of FANCD2 (Wang et al. 2007).

FA-F is reported in less than 5% of patients (Kennedy and D'Andrea 2005). FANCF is 42kDA (Wang 2007), a member of the FA core complex, and is required to stabilize interaction of the other core complex members (Leveille et al. 2004). FANCF has a region that is homologous with the prokaryotic RNA-binding protein, ROM. However, the homology is not within the RNA-binding motif of ROM and overall the significance of the homology is unknown as mutation at this motif does not impair the function of FANCF within the FA pathway (de Winter et al. 2000).

FA-G accounts for 10% of FA patients (Kennedy and D'Andrea 2005). FANCG is 68 kDa and is also known as XRCC9, originally identified as a UV resistance factor in chinese hamster ovary cells (Liu et al. 1997; de Winter et al. 1998). Like other FA proteins, FANCG has no recognizable functional motifs or domains, but does possess several tetratricopeptide (TPR) motifs, often involved in mediating protein complex interactions (Blom et al. 2004). The presence of these motifs is consistent with the participation of FANCG in the FA core complex (Blom et al. 2004; Wang 2007). FANCG has several known sites of phosphorylation; serine 7, serine 383, and serine 387 (Mi et al. 2004; Qiao et al. 2004). Phosphorylation at serines 383 and 387 occurs during mitosis and is thought to help direct localization of the FA core complex away from the condensed chromatin (Qiao et al. 2001; Mi et al. 2004). FANCG is known to participate in the FA core complex as well as function outside of the FA pathway. Phosphorylation of FANCG at serine 7 leads to its participation in a protein complex that includes FANCD1/BRCA2 and Rad51, but the function of this complex lies outside of the known FA pathway (Hussain et al. 2003), as it is independent of upstream members of the FA core complex for its formation.

FA-I is rare among known FA patients (Kennedy and D'Andrea 2005). The recent identification of the gene responsible for FA-I, FANCI, was accompanied by the discovery that it possesses sequence and functional homology to FANCD2 (Dorsman et al. 2007; Meijer 2007; Sims et al. 2007; Smogorzewska et al. 2007). Like FANCD2, FANCI is monoubiquitylated upon DNA damage or S-phase, in this case at lysine 523, which elevates its molecular weight from 140 to 147 kDa (Wang 2007). FANCI and FANCD2 bind one another in a dimeric or multimeric complex and the functions of each protein depends on their interaction (Sims et al. 2007; Smogorzewska et al. 2007).

FANCJ, accounts for less than 2% of FA cases (Kennedy and D'Andrea 2005) and approximately 0.2% of familial breast cancer cases (Seal et al. 2006). These disorders are caused by mutations in FANCJ, also known as *BRCA*1-associated C-terminal helicase (BACH1) or BRIP1. FANCJ/BACH1 is a 140 kDa DEAH-box containing helicase that unwinds DNA in a 5' to 3' direction (Cantor et al. 2004; Gupta et al. 2005) unlike other similar helicases that unwind in a 3' to 5' direction, such as BLM or WRN, which are also associated with disorders of DNA repair. Interestingly, one of the most common mutations in FANCJ leads to premature truncation of the protein at arginine 798 (R798X). When this mutation is found on only one allele, then the patient is susceptible to familial breast cancer (Cantor et al. 2004). However, when a patient is homozygous for this mutation, the result is FA (Levran et al. 2005).

FA-L accounts for less than five percent of all FA cases reported. The gene, FANCL, codes for the E3 ubiquitin ligase PHF9. The protein is 43 kDa (Wang 2007), has a PHD finger-type E3 ubiquitin ligase motif, and has been shown to have autopolyubiquitylating activity. FANCL is a member of the FA core complex and is required for monoubiquitylation of FANCD2, but has not been shown to directly ubiquitylate FANCD2 in vitro or in vivo. While an E3 ubiquitin ligase is responsible for transferring ubiquitin to a target protein, an E2 subunit, known as a conjugating enzyme, presents ubiquitin to the E3. The E2 subunit UBE2T has been shown to interact directly with FANCL, and UBE2T is necessary for FANCD2 monoubiquitination *in vivo* (Machida et al. 2006). Interestingly, no FA patients have been identified with a UBE2T mutation.

FA-M is extremely rare. This FA subtype is a result of a mutation in the causative gene, FANCM, which encodes a 250 kDa protein (Meetei et al. 2005). FANCM is the most conserved of all the FA proteins, with homologs in vertebrates, invertebrates, yeast, and an ortholog in archaebacteria called HEF1 (Wang 2007). Among the FA proteins, FANCM is exceptional in that it contains two conserved, recognizable motifs, a helicase motif near the N-terminus, and an endonuclease motif near the C-terminus that shares a degenerate homology to the excision repair protein ERCC4/XPF. The degenerate sequence of the endonuclease domain likely renders it nonfunctional. Though neither helicase nor endonuclease activity have been shown, FANCM is capable of DNA translocase activity (Meetei et al. 2005) and DNA binding (Mosedale et al. 2005). FANCM is hyperphosphorylated in response to DNA damage (Meetei et al. 2005), and is thought to facilitate binding of the FA core complex to chromatin. In addition, FANCM also requires a cofactor for its function, FAAP24, whose function is unknown (Ciccia et al. 2007). No FA patients have been identified with mutations in FAAP24 (Ciccia et al. 2007).

FA-N, a 140 kDa protein (Wang 2007), accounts for approximately 2% of all reported cases of FA (Kennedy and D'Andrea 2005) and mutation of the responsible gene FANCN/PALB2 results in 1.1% of familial breast cancer patients (Rahman et al. 2007). PALB2 (partner and localizer of BRCA2)/FANCN colocalizes with BRCA2/FANCD1 and promotes its participation in DNA repair complexes (Rahman et al. 2007). Interestingly, FA-N and FA-D1 patients differ from other FA patients in that they are susceptible to Wilms tumor and medulloblastoma, which are extremely rare in the general FA population (Reid et al. 2005; Taniguchi and D'Andrea 2006; Reid et al. 2007).

BRCA1

The discovery of BRCA1 was published in 1990 (Hall et al. 1990) and found in 52% of familial breast cancer pedigrees, that is, families with at least four cases of breast cancer (Ford et al. 1998). BRCA1 is regarded as a classical tumor suppressor gene (Boulton 2006). BRCA1 patients have monoallelic mutations, and cancer generally arises following LOH and loss of the wild-type allele (Smith et al. 1994, Kelsell et al. 1993). BRCA1 gene encodes a 204 kDa, 1,863aa protein with a highly conserved RING finger motif at the N-terminus, and two BRCT repeats at its C-terminus. The functions of BRCA1 identified thus far center on repair of DNA damage, transcription and promoter binding, and cell cycle arrest. Thus far, BRCA1 has been identified with E3 ubiquitin ligase activity (Hashizume et al. 2001; Ruffner et al. 2001), cell cycle checkpoint arrest (Xu et al. 2001), chromatin structure management (Turner et al. 2004), regulation of other DNA repair factors (Kastan and Bartek 2004), and participation in DNA repair protein complexes (Scully et al. 1997; Chen et al. 1998). BRCA1 has been shown to immunoprecipitate with monoubiquitylated FANCD2 (Garcia-Higuera et al. 2001; Vandenberg et al. 2003). Furthermore, BRCA1 colocalizes with FANCD2 in DNA repair foci and knockdown of BRCA1 expression leads to chromosomal instability in a manner comparable to the instability seen with loss of FANCD2 expression. These data strongly implicate BRCA1 as a key participant in the FA-BRCA pathway.

Clinical Observations

Patients with FA have biallelic mutations and are vulnerable to both hematological malignancies and solid tumors (Joenje and Patel 2001; Kennedy and D'Andrea 2005). The eight members of the FA core complex do not confer susceptibility to breast cancer in the heterozygous or homozygous state, although FANCF silencing via methylation of the promoter has been seen in a high percentage of sporadic ovarian carcinomas (see below). All FA genes that overlap with breast cancer susceptibility appear to function downstream of the monoubiquitination of FANCD2. FANCD1/BRCA2, FANCN/PALB2, and FANCJ/BACH1 are unique in that monoallelic mutation leads to familial breast cancer, whereas biallelic mutation results in FA (Cantor and Andreassen 2006; Rahman et al. 2007). FANCJ/BACH1 patients with the mutation that causes premature truncation (R798X) on only one allele are susceptible to breast cancer, whereas those with the same mutation on both alleles are born with FA (Levran et al. 2005). Similarly, FANCN/PALB2 patients with nontruncating variants are found in families with breast cancer, whereas all FA-N patients have biallelic mutations that resulted in premature truncation of the FANCN protein (Rahman et al. 2007).

Animal studies have shown biallelic elimination of either BRCA gene is embryonic lethal and patients with familial mutations of BRCA1 have only one mutant allele. In BRCA1 patients, cancer arises either because of a dominant mutant allele or loss of heterozygosity and loss of the wild-type allele (Claus et al. 1991). Patients with familial mutations in FANCD1/BRCA2 express a partially active mutant form of this protein, typically resulting in a hypomorphic phenotype, and cancer results from loss of heterozygosity. FA-D1 patients have biallelic mutations that result in expression of a truncated protein (Howlett et al. 2002).

FA proteins have been implicated in sporadic tumorigenesis. Epigenetic silencing of FANCF has been seen in 17% of sporadic breast cancers, 21% of ovarian cancers, and 30% of cervical cancers (Turner et al. 2004). Epigenetic silencing and/or mutation of the FANCC and FANCG genes were found in cases of young-onset pancreatic cancer (van der Heijden et al. 2003; Couch et al. 2005). Silencing of the FA genes occurred when the DNA was methylated at cystine residues, which attracts binding proteins that make the gene sequence inaccessible to transcription factors.

DNA Repair and Translesion Synthesis

Repair of ICL occurs in coordination with DNA replication by translesion DNA synthesis, homologous recombination, or a combination of the two. When a DNA replication fork encounters an ICL, the DNA strands cannot be separated and the DNA polymerase machinery stalls. Stalled replication results in exposure of single-stranded DNA, which attracts and binds RPA. The accumulation of RPA at the DNA site activates the ATR-interacting protein, ATRIP. The ATRIP/ATR complex proceeds to phosphorylate a series of factors including FANCD2, BRCA2/FANCD1, and BRCA1. Shortly thereafter, one of the replication polymerase factors, PCNA, becomes monoubiquitylated. The monoubiquitylation of PCNA causes ejection of the high-fidelity polymerase α and recruitment of the lower-fidelity translesion (TLS) DNA polymerases. The TLS polymerases are able to synthesize short stretches of DNA using the damaged region as a template. Because of their ability to use damaged DNA as a template, the TLS polymerases are prone to errors. The TLS polymerases also have low processivity and are thus ejected from the DNA after relatively few base additions. The deubiquitylating enzyme, USP1, removes the ubiquitin moiety on PCNA whereupon the high-fidelity DNA polymerase α is recruited to continue normal DNA replication. Alternatively, the damaged DNA can be repaired with homologous recombination (HR). In HR, the cell uses the undamaged sister chromatid as a template to conduct sequence repair on a damaged DNA site. Here, the ICL is clipped out of one strand, leaving a gap in that strand. The presence of the replication fork and nicked DNA leads to a single strand being exposed. RAD51 binds to the single strand and facilitates its invasion into a homologous region of the sister chromatid, creating a D-loop. In vitro, the binding of RAD51 to a single-stranded DNA oligonucleotide is sufficient to facilitate its invasion into homologous double-stranded DNA. However, in vivo, RAD51 is known to interact with BRCA1, FANCD1/BRCA2, FANCD2, and other DNA repair factors, although the mechanisms and necessity of these interactions remains unknown. Once the single-strand DNA has invaded its sister chromatid, the RAD51 is dissociated, and polymerases extend the strand using the sister chromatid as a template. After extension, the invading strand is ejected, returning to its damaged partner and acting as a template for final clipping of the ICL lesion and sequence repair. Growing evidence suggests that the FA/BRCA pathway plays multiple, critical roles in both HR and TLS processes.

Therapeutic Relevance

In addition to the genetic syndrome of FA and its accompanying constellation of clinical signs and symptoms, the BRCA/FA pathway has clear implications in the general world of oncology. Naturally, on the basis of classic tumor suppressor gene literature, such as p53, the hypothesis was that FA heterozygotes should display an increase in cancer. To date, such data have not been put forward, except for the obvious cases of familial breast cancer, as in BRCA2/FANCD1, BACH1FANCJ, and PALB1/FANCN. Previous to the identification of these familial breast cancer genes as FA genes, there was not link of FA heterozyogosity to increased cancer risk because of the rarity of these FA gene types compared to the FANCA, FANCC, and FANCG groups which comprise the large majority of all FA complementation groups.

AMLs quite naturally have also been screened for abnormalities in the FA pathway, with somatic deletion mutations noted in FANCA. These sporadic AMLs have variably been noted to be either chemotherapy sensitive or resistant, at least *in vitro*, which is probably testimony to the genomic instability that the FA pathway can engender, resulting in growth advantage potential.

Other sporadic tumors have been noted to have defects in FA protein expression, most notably involving FANCF. These aberrations have been explained by methylation of the FANCF promoter rather than acquired or germline mutations in the coding region of the gene. Crosslinker-sensitive ovarian cell lines display a subset of affected FANCF promoters and FANCF has been also implicated in head and neck, and lung cancers. In addition, decreased expression of FANCA, FANCC, and FANCG have been associated with pancreatic cancers accompanied by diminished FANCD2 foci formation, which implies that the FA pathway is indeed functionally impaired. Interestingly, pancreatic cancers are part of the constellation of tumors seen in familial breast cancers, in which BRCA2/FANCD2, PALB2/FANCN, and BACH1/FANCJ are examples, yet cells with these mutant backgrounds still display FANCD2 foci.

On the other hand, the practical effect on clinical significance is less clearly understood. While some have noted an association of poor survival with diminished FANCF expressing tumors, other studies have reported that no apparent difference exists in patient's long-term survival. A similar phenomenon exists in familial breast cancers where clinical significance is uncertain. Even though cells mutant and/or knocked out for BRCA genes in cell culture can show marked DNA damage sensitivity, this does not necessarily correlate to increased tumor formation. In addition, survival of patients with BRCA tumors has both been shown to be diminished or unchanged vs. non-BRCA tumors.

The therapeutic approach to the FA/BRCA pathway has focused on two avenues. The first approach takes advantage of the mutant state of BRCA-associated tumors. For example, BRCA1 binds to and may regulate HDAC and HATs. Use of inhibitors of chromatin modifications such as those regulated through BRCA1 have been a current avenue of clinical trials, although such efforts have been largely unimpressive. A second approach has been to target the FA/BRCA pathway in order to specifically render cells sensitive to ICLs. Such an approach would still present the problem that all chemotherapy approaches have: toxicity to normal tissues, although presumably cancers that are proproliferative could be effectively targeted. On the other hand, a reasonably large number of head and neck tumors and ovarian tumors have been shown to exhibit diminished FA protein levels, and such tumors could be prime candidates for ICL-based regimens. Screening for FA pathway status could reasonably be part of these tumors' pathologic workup.

References

(1996). "Positional cloning of the Fanconi anaemia group A gene. The Fanconi anaemia/breast cancer consortium." Nat Genet **14**(3): 324–8.

Abu-Issa, R., G. Eichele, et al. (1999). "Expression of the Fanconi anemia group A gene (Fanca) during mouse embryogenesis." Blood **94**(2): 818–24.

Alter, B. (2003). Inherited bone marrow failure syndromes. In Orkin, S., Nathan, DG., Ginsburg, D., Look, T. (eds.), "Hematology of Infancy and Childhood, 6th ed." WB Saunders, Philadelphia, PA, pp. 280–365.

Alter, BP., N. Young (1993). The bone marrow failure syndromes. In Oski, FA., Nathan, DG. (eds.), "Hematology of Infancy and Childhood." WB Saunders, Philadelphia, PA, pp. 216–316.

Andreassen, PR., AD. D'Andrea, et al. (2004). "ATR couples FANCD2 monoubiquitination to the DNA-damage response." Genes Dev **18**(16): 1958–63.

Auerbach A., M. Buchwald, H. Joenje (1997). Fanconi anemia. In Vogelstein, B., Kinzler. KW (eds.), "Genetics of Cancer." McGraw-Hill, New York, NY, pp. 317–32.

Blom, E., HJ. van de Vrugt, et al. (2004). "Multiple TPR motifs characterize the Fanconi anemia FANCG protein." DNA Repair (Amst) **3**(1): 77–84.

Bogliolo, M., A. Lyakhovich, et al. (2007). "Histone H2AX and Fanconi anemia FANCD2 function in the same pathway to maintain chromosome stability." EMBO J **26**: 1340–51.

Boulton, SJ. (2006). "Cellular functions of the BRCA tumour-suppressor proteins." Biochem Soc Trans **34**(Pt 5): 633–45.

Broxmeyer, HE., DE. Williams, et al. (1986). "The suppressive influences of human tumor necrosis factors on bone marrow hematopoietic progenitor cells from normal donors and patients with leukemia: synergism of tumor necrosis factor and interferon-gamma." J Immunol **136**(12): 4487–95.

Bryder, D., V. Ramsfjell, et al. (2001). "Self-renewal of multipotent long-term repopulating hematopoietic stem cells is negatively regulated by Fas and tumor necrosis factor receptor activation." J Exp Med **194**(7): 941–52.

Cantor, S., R. Drapkin, et al. (2004). "The BRCA1-associated protein BACH1 is a DNA helicase targeted by clinically relevant inactivating mutations." Proc Natl Acad Sci U S A **101**(8): 2357–62.

Cantor, SB., PR. Andreassen (2006). "Assessing the link between BACH1 and BRCA1 in the FA pathway." Cell Cycle **5**(2): 164–7.

Chen, G., DV. Goeddel (2002). "TNF-R1 signaling: a beautiful pathway." Science **296**(5573): 1634–5.

Chen, J., DP. Silver, et al. (1998). "Stable interaction between the products of the BRCA1 and BRCA2 tumor suppressor genes in mitotic and meiotic cells." Mol Cell **2**(3): 317–28.

Christianson, TA., GC. Bagby (2000). "FANCA protein binds FANCG proteins in an intracellular complex." Blood **95**(2): 725–6.

Ciccia, A., C. Ling, et al. (2007). "Identification of FAAP24, a Fanconi anemia core complex protein that interacts with FANCM." Mol Cell **25**(3): 331–43.

Claus, EB., N. Risch, et al. (1991). "Genetic analysis of breast cancer in the cancer and steroid hormone study." Am J Hum Genet **48**(2): 232–42.

Collins, N., GM. Kupfer (2005). "Molecular pathogenesis of Fanconi anemia." Int J Hematol **82**(3): 176–83.

Couch, FJ., MR. Johnson, et al. (2005). "Germ line Fanconi anemia complementation group C mutations and pancreatic cancer." Cancer Res **65**(2): 383–6.

De Silva, IU., PJ. McHugh, et al. (2000). "Defining the roles of nucleotide excision repair and recombination in the repair of DNA interstrand cross-links in mammalian cells." Mol Cell Biol **20**(21): 7980–90.

de Winter, JP., F. Leveille, et al. (2000). "Isolation of a cDNA representing the Fanconi anemia complementation group E gene." Am J Hum Genet **67**(5): 1306–8.

de Winter, JP., L. van der Weel, et al. (2000). "The Fanconi anemia protein FANCF forms a nuclear complex with FANCA, FANCC and FANCG." Hum Mol Genet **9**(18): 2665–74.

de Winter, JP., Q. Waisfisz, et al. (1998). "The Fanconi anaemia group G gene FANCG is identical with XRCC9." Nat Genet **20**(3): 281–3.

Dorsman, JC., M. Levitus, et al. (2007). "Identification of the Fanconi anemia complementation group I gene, FANCI." Cell Oncol **29**(3): 211–8.

Dufour, C., A. Corcione, et al. (2003). "TNF-alpha and IFN-gamma are overexpressed in the bone marrow of Fanconi anemia patients and TNF-alpha suppresses erythropoiesis in vitro." Blood **102**(6): 2053–9.

Dunn, J., M. Potter, et al. (2006). "Activation of the Fanconi anemia/BRCA pathway and recombination repair in the cellular response to solar ultraviolet light." Cancer Res **66**(23): 11140–7.

Dybedal, I., D. Bryder, et al. (2001). "Tumor necrosis factor (TNF)-mediated activation of the p55 TNF receptor negatively regulates maintenance of cycling reconstituting human hematopoietic stem cells." Blood **98**(6): 1782–91.

Fagerlie, SR., GC. Bagby (2006). "Immune defects in Fanconi anemia." Crit Rev Immunol **26**(1): 81–96.

Fanconi, G. (1967). "Familial constitutional panmyelocytopathy, Fanconi's anemia (F.A.). I. Clinical aspects." Semin Hematol **4**(3): 233–40.

Foe, JR., MA. Rooimans, et al. (1996). "Expression cloning of a cDNA for the major Fanconi anaemia gene, FAA." Nat Genet **14**(4): 488.

Ford, D., DF. Easton, et al. (1998). "Genetic heterogeneity and penetrance analysis of the BRCA1 and BRCA2 genes in breast cancer families. The Breast Cancer Linkage Consortium." Am J Hum Genet **62**(3): 676–89.

Futaki, M., T. Igarashi, et al. (2002). "The FANCG Fanconi anemia protein interacts with CYP2E1: possible role in protection against oxidative DNA damage." Carcinogenesis **23**(1): 67–72.

Garcia-Higuera, I., Y. Kuang, et al. (1999). "Fanconi anemia proteins FANCA, FANCC, and FANCG/XRCC9 interact in a functional nuclear complex." Mol Cell Biol **19**(7): 4866–73.

Garcia-Higuera, I., T. Taniguchi, et al. (2001). "Interaction of the Fanconi anemia proteins and BRCA1 in a common pathway." Mol Cell **7**(2): 249–62.

Gupta, R., S. Sharma, et al. (2005). "Analysis of the DNA substrate specificity of the human BACH1 helicase associated with breast cancer." J Biol Chem **280**(27): 25450–60.

Hall, JM., MK. Lee, et al. (1990). "Linkage of early-onset familial breast cancer to chromosome 17q21." Science **250**(4988): 1684–9.

Hashizume, R., M. Fukuda, et al. (2001). "The RING heterodimer BRCA1-BARD1 is a ubiquitin ligase inactivated by a breast cancer-derived mutation." J Biol Chem 276(18): 14537–40.

Hejna, JA., CD. Timmers, et al. (2000). "Localization of the Fanconi anemia complementation group D gene to a 200-kb region on chromosome 3p25.3." Am J Hum Genet 66(5): 1540–51.

Ho, GP., S. Margossian, et al. (2006). "Phosphorylation of FANCD2 on two novel sites is required for mitomycin C resistance." Mol Cell Biol 26(18): 7005–15.

Houghtaling, S., A. Newell, et al. (2005). "Fancd2 functions in a double strand break repair pathway that is distinct from non-homologous end joining." Hum Mol Genet 14(20): 3027–33.

Houghtaling, S., C. Timmers, et al. (2003). "Epithelial cancer in Fanconi anemia complementation group D2 (Fancd2) knockout mice." Genes Dev 17(16): 2021–35.

Howlett, NG., T. Taniguchi, et al. (2002). "Biallelic inactivation of BRCA2 in Fanconi anemia." Science 297(5581): 606–9.

Hussain, S., E. Witt, et al. (2003). "Direct interaction of the Fanconi anaemia protein FANCG with BRCA2/FANCD1." Hum Mol Genet 12(19): 2503–10.

Joenje, H., KJ. Patel (2001). "The emerging genetic and molecular basis of Fanconi anaemia." Nat Rev Genet 2(6): 446–57.

Kastan, MB., J. Bartek (2004). "Cell-cycle checkpoints and cancer." Nature 432(7015): 316–23.

Kelsell, DP., DM. Black, et al. (1993). "Genetic analysis of the BRCA1 region in a large breast/ovarian family: refinement of the minimal region containing BRCA1" Hum Mol Genet 2(11): 1823–8.

Kennedy, RD., AD. D'Andrea (2005). "The Fanconi anemia/BRCA pathway: new faces in the crowd." Genes Dev 19(24): 2925–40.

Kitao, H., K. Yamamoto, et al. (2006). "Functional interplay between BRCA2/FancD1 and FancC in DNA repair." J Biol Chem 281(30): 21312–20.

Kupfer, GM., D. Naf, et al. (1997). "The Fanconi anaemia proteins, FAA and FAC, interact to form a nuclear complex." Nat Genet 17(4): 487–90.

Leveille, F., E. Blom, et al. (2004). "The Fanconi anemia gene product FANCF is a flexible adaptor protein." J Biol Chem 279(38): 39421–30.

Levitus, M., MA. Rooimans, et al. (2004). "Heterogeneity in Fanconi anemia: evidence for 2 new genetic subtypes." Blood 103(7): 2498–503.

Levran, O., C. Attwooll, et al. (2005). "The BRCA1-interacting helicase BRIP1 is deficient in Fanconi anemia." Nat Genet 37(9): 931–3.

Liu, N., JE. Lamerdin, et al. (1997). "The human XRCC9 gene corrects chromosomal instability and mutagen sensitivities in CHO UV40 cells." Proc Natl Acad Sci U S A 94(17): 9232–7.

Lo Ten Foe, JR., MA. Rooimans, et al. (1996). "Expression cloning of a cDNA for the major Fanconi anaemia gene, FAA." Nat Genet 14(3): 320–3.

Machida, YJ., Y. Machida, et al. (2006). "UBE2T is the E2 in the Fanconi anemia pathway and undergoes negative autoregulation." Mol Cell 23(4): 589–96.

Meetei, AR., JP. de Winter, et al. (2003). "A novel ubiquitin ligase is deficient in Fanconi anemia." Nat Genet 35(2): 165–70.

Meetei, AR., M. Levitus, et al. (2004). "X-linked inheritance of Fanconi anemia complementation group B." Nat Genet 36(11): 1219–24.

Meetei, AR., AL. Medhurst, et al. (2005). "A human ortholog of archaeal DNA repair protein Hef is defective in Fanconi anemia complementation group M." Nat Genet 37(9): 958–63.

Meetei, AR., Z. Yan, et al. (2004). "FANCL replaces BRCA1 as the likely ubiquitin ligase responsible for FANCD2 monoubiquitination." Cell Cycle 3(2): 179–81.

Meijer, GA. (2007). "The 13th Fanconi anemia gene identified: FANCI–importance of the 'Fanconi anemia pathway' for cellular oncology." Cell Oncol 29(3): 181–2.

Mi, J., F. Qiao, et al. (2004). "FANCG is phosphorylated at serines 383 and 387 during mitosis." Mol Cell Biol 24(19): 8576–85.

Miki, Y., J. Swensen, et al. (1994). "A strong candidate for the breast and ovarian cancer susceptibility gene BRCA1." Science 266(5182): 66–71.

Mosedale, G., W. Niedzwiedz, et al. (2005). "The vertebrate Hef ortholog is a component of the Fanconi anemia tumor-suppressor pathway." Nat Struct Mol Biol 12(9): 763–71.

Murase, T., T. Hotta, et al. (1987). "Effect of recombinant human tumor necrosis factor on the colony growth of human leukemia progenitor cells and normal hematopoietic progenitor cells." Blood **69**(2): 467–72.

Naf, D., GM. Kupfer, et al. (1998). "Functional activity of the fanconi anemia protein FAA requires FAC binding and nuclear localization." Mol Cell Biol **18**(10): 5952–60.

Narod, SA. (2002). "Modifiers of risk of hereditary breast and ovarian cancer." Nat Rev Cancer **2**(2): 113–23.

Narod, SA., WD. Foulkes (2004). "BRCA1 and BRCA2: 1994 and beyond." Nat Rev Cancer **4**(9): 665–76.

Niedernhofer, LJ., H. Odijk, et al. (2004). "The structure-specific endonuclease Ercc1-Xpf is required to resolve DNA interstrand cross-link-induced double-strand breaks." Mol Cell Biol **24**(13): 5776–87.

Pace, P., M. Johnson, et al. (2002). "FANCE: the link between Fanconi anaemia complex assembly and activity." Embo J **21**(13): 3414–23.

Pagano, G., H. Youssoufian (2003). "Fanconi anaemia proteins: major roles in cell protection against oxidative damage." Bioessays **25**(6): 589–95.

Pang, Q., TA. Christianson, et al. (2001). "The Fanconi anemia complementation group C gene product: structural evidence of multifunctionality." Blood **98**(5): 1392–401.

Qiao, F., J. Mi, et al. (2004). "Phosphorylation of fanconi anemia (FA) complementation group G protein, FANCG, at serine 7 is important for function of the FA pathway." J Biol Chem **279**(44): 46035–45.

Qiao, F., A. Moss, et al. (2001). "Fanconi anemia proteins localize to chromatin and the nuclear matrix in a DNA damage- and cell cycle-regulated manner." J Biol Chem **276**(26): 23391–6.

Rahman, N., S. Seal, et al. (2007). "PALB2, which encodes a BRCA2-interacting protein, is a breast cancer susceptibility gene." Nat Genet **39**(2): 165–7.

Rathbun, RK., GR. Faulkner, et al. (1997). "Inactivation of the Fanconi anemia group C gene augments interferon-gamma-induced apoptotic responses in hematopoietic cells." Blood **90**(3): 974–85.

Rebbeck, TR., FJ. Couch, et al. (1996). "Genetic heterogeneity in hereditary breast cancer: role of BRCA1 and BRCA2." Am J Hum Genet **59**(3): 547–53.

Reid, S., A. Renwick, et al. (2005). "Biallelic BRCA2 mutations are associated with multiple malignancies in childhood including familial Wilms tumour." J Med Genet **42**(2): 147–51.

Reid, S., D. Schindler, et al. (2007). "Biallelic mutations in PALB2 cause Fanconi anemia subtype FA-N and predispose to childhood cancer." Nat Genet **39**(2): 162–4.

Rosselli, F., J. Sanceau, et al. (1994). "Abnormal lymphokine production: a novel feature of the genetic disease Fanconi anemia. II. In vitro and in vivo spontaneous overproduction of tumor necrosis factor alpha." Blood **83**(5): 1216–25.

Ruffner, H., CA. Joazeiro, et al. (2001). "Cancer-predisposing mutations within the RING domain of BRCA1: loss of ubiquitin protein ligase activity and protection from radiation hypersensitivity." Proc Natl Acad Sci U S A **98**(9): 5134–9.

Schultz, JC., NT. Shahidi (1993). "Tumor necrosis factor-alpha overproduction in Fanconi's anemia." Am J Hematol **42**(2): 196–201.

Scully, R., J. Chen, et al. (1997). "Association of BRCA1 with Rad51 in mitotic and meiotic cells." Cell **88**(2): 265–75.

Seal, S., D. Thompson, et al. (2006). "Truncating mutations in the Fanconi anemia J gene BRIP1 are low-penetrance breast cancer susceptibility alleles." Nat Genet **38**: 1239–41.

Siddique, MA., K. Nakanishi, et al. (2001). "Function of the Fanconi anemia pathway in Fanconi anemia complementation group F and D1 cells." Exp Hematol **29**(12): 1448–55.

Sims, AE., E. Spiteri, et al. (2007). "FANCI is a second monoubiquitinated member of the Fanconi anemia pathway." Nat Struct Mol Biol **14**(6): 564–7.

Smith, SA., RA. DiCioccio, et al. (1994). "Localisation of the breast-ovarian cancer susceptibility gene (BRCA1) on 17q12–21 to an interval of <or=1 CM". Genes Chromosomes Cancer **10**(1): 71–6.

Smogorzewska, A., S. Matsuoka, et al. (2007). "Identification of the FANCI protein, a monoubiq-uitinated FANCD2 paralog required for DNA repair." Cell **129**(2): 289–301.

Sonoda E., MS. Sasaki, et al. (1999). "3-Sister chromatid exchanges are mediated by homologous recombination in vertebrate cells." Mol Cell Biol **19**(7): 5166–9.

Strathdee, CA., H. Gavish, et al. (1992). "Cloning of cDNAs for Fanconi's anaemia by functional complementation." Nature **358**(6385): 434.

Taniguchi, T., AD. D'Andrea (2006). "Molecular pathogenesis of Fanconi anemia: recent progress." Blood **107**(11): 4223–33.

Taniguchi, T., I. Garcia-Higuera, et al. (2002). "Convergence of the fanconi anemia and ataxia telangiectasia signaling pathways." Cell **109**(4): 459–72.

Thomashevski, A., AA. High, et al. (2004). "The Fanconi anemia core complex forms four com-plexes of different sizes in different subcellular compartments." J Biol Chem **279**(25): 26201–9.

Timmers, C., T. Taniguchi, et al. (2001). "Positional cloning of a novel Fanconi anemia gene, FANCD2." Mol Cell **7**(2): 241–8.

Turner, JM., O. Aprelikova, et al. (2004). "BRCA1, histone H2AX phosphorylation, and male meiotic sex chromosome inactivation." Curr Biol **14**(23): 2135–42.

Turner, N., A. Tutt, et al. (2004). "Hallmarks of 'BRCAness' in sporadic cancers." Nat Rev Cancer **4**(10): 814–9.

van der Heijden, MS., CJ. Yeo, et al. (2003). "Fanconi anemia gene mutations in young-onset pancreatic cancer." Cancer Res **63**(10): 2585–8.

Vandenberg, CJ., F. Gergely, et al. (2003). "BRCA1-independent ubiquitination of FANCD2." Mol Cell **12**(1): 247–54.

Waisfisz, Q., JP. de Winter, et al. (1999). "A physical complex of the Fanconi anemia proteins FANCG/XRCC9 and FANCA." Proc Natl Acad Sci U S A **96**(18): 10320–5.

Wajant, H., K. Pfizenmaier, et al. (2003). "Tumor necrosis factor signaling." Cell Death Differ **10**(1): 45–65.

Wang, W. (2007). "Emergence of a DNA-damage response network consisting of Fanconi anae-mia and BRCA proteins." Nat Rev Genet **8**(10): 735–48.

Wang, X., PR. Andreassen, et al. (2004). "Functional interaction of monoubiquitinated FANCD2 and BRCA2/FANCD1 in chromatin." Mol Cell Biol **24**(13): 5850–62.

Wang, X., RD. Kennedy, et al. (2007). "Chk1-mediated phosphorylation of FANCE is required for the Fanconi anemia/BRCA pathway." Mol Cell Biol **27**(8): 3098–108.

Whitney, M., M. Thayer, et al. (1995). "Microcell mediated chromosome transfer maps the Fanconi anaemia group D gene to chromosome 3p." Nat Genet **11**(3): 341–3.

Whitney, MA., G. Royle, et al. (1996). "Germ cell defects and hematopoietic hypersensitivity to gamma-interferon in mice with a targeted disruption of the Fanconi anemia C gene." Blood **88**(1): 49–58.

Wooster, R., G. Bignell, et al. (1995). "Identification of the breast cancer susceptibility gene BRCA2." Nature **378**(6559): 789–92.

Wooster, R., SL. Neuhausen, et al. (1994). "Localization of a breast cancer susceptibility gene, BRCA2, to chromosome 13q12-13." Science **265**(5181): 2088–90.

Xu, B., S. Kim, et al. (2001). "Involvement of Brca1 in S-phase and G(2)-phase checkpoints after ionizing irradiation." Mol Cell Biol **21**(10): 3445–50.

Yamashita, T., GM. Kupfer, et al. (1998). "The fanconi anemia pathway requires FAA phosphoryla-tion and FAA/FAC nuclear accumulation." Proc Natl Acad Sci U S A **95**(22): 13085–90.

Yamashita, T., N. Wu, et al. (1996). "Clinical variability of Fanconi anemia (type C) results from expression of an amino terminal truncated Fanconi anemia complementation group C polypeptide with partial activity." Blood **87**(10): 4424–32.

Targeting the Sumoylation Pathway

Pooja Pungaliya and Eric Rubin

Abstract Sumoylation is a dynamic process where SUMO (small ubiquitin-like modifier) is covalently conjugated and deconjugated from target proteins to regulate their cellular localization, stability, and function. Analogous to the ubiquitination pathway, sumoylation involves E1, E2, and E3 enzymes that are engaged in the processing, ligation, and recycling of SUMO. Sumoylation influences proteins involved in various cellular processes including stress and transcription, with dysregulation of these pathways leading to biological dysfunction and disease. A number of oncogenes and tumor suppressors involved in genomic integrity and chromatin remodeling have been identified as sumoylation targets. Because variations in the sumoylation pathway lead to diseases such as cancer, enzymes involved in the pathway serve as potential therapeutic targets.

Keywords SUMO • E3 ligase • Ubc9 • Isopeptidase • Chromatin • Transcriptional regulation • Tumorigenesis • Neurodegeneration

Introduction: SUMO Proteins and Conjugation Enzymes

Posttranslational protein modification by ubiquitin and ubiquitin-like proteins such as SUMO (small ubiquitin-like modifier) are increasingly recognized as important in human disease. This chapter will review current knowledge regarding components of cellular sumoylation pathways, with an emphasis on potential therapeutic targets.

SUMO protein orthologs are found in all eukaryotes, with three paralogs identified in humans (SUMO-1, SUMO-2, and SUMO-3). These 92–97 amino acid proteins share limited sequence homology with ubiquitin, but have a C-terminal structure that is very similar to that of ubiquitin (Bayer et al. 1998). SUMO-2 and SUMO-3

P. Pungaliya
Postdoctoral Fellow, Biological Technologies, Wyeth Research, 87 Cambridge Park Drive, Cambridge, MA 02140, 617.665.8341
e-mail: pungalp@wyeth.com

are ~95% similar in sequence, whereas SUMO-1 shares only about 50% homology with these isoforms. SUMO isoforms are attached to the ε-amino group of substrate lysines via an isopeptide bond involving a C-terminal diglycine sequence of SUMO. The substrate lysine residue is often within a sequence comprised of ΨKXE, where Ψ is a hydrophobic amino acid (Rodriguez et al. 2001; Sampson et al. 2001). As discussed below, accumulating evidence supports a distinct cellular role for SUMO-1 versus SUMO-2/3 isoforms. Protein modification by SUMO is regulated via conjugating and deconjugating enzymes, with these processes also controlled by phosphorylation of either the substrate or perhaps proteins involved in the conjugating (Hecker et al. 2006) or deconjugating processes. Similar to ubiquitin, SUMO attachment to target proteins involves 3 enzymes: an activating enzyme (E1), conjugating enzyme (E2), and ligase (E3) (Fig. 1). Furthermore, analogous to the ubiquitin

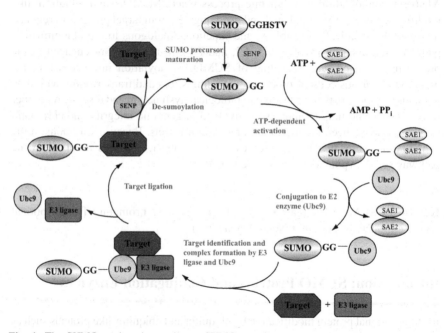

Fig. 1 The SUMO conjugating pathway. SUMO proteins are precursors that undergo posttranslational cleavage by SUMO hydrolases (SENP). Maturation of SUMO exposes the C-terminal di-glycine motif that is required for conjugation. Next, SUMO is conjugated to the SUMO E1 activating enzyme SAE1/SAE2 in an ATP-dependent manner followed by subsequent conjugation to the SUMO E2 conjugating enzyme Ubc9. Then, an E3 ligase through scaffolding brings Ubc9 and the target protein within close proximity to allow transfer of SUMO to the target protein where an isopeptide bond is formed. Finally, the SUMO conjugate on the target protein is cleaved by a SUMO protease and the SUMO protein is recycled.

Table 1 Components of the mammalian sumoylation pathway

SUMO isoforms	E1 enzyme	E2 enzyme	E3 enzymes	Proteases
SUMO-1	SAE-1/SAE-2	UBC-9	PIAS-1–4 RANBP-2	Ulp1-like (SENP1-3,
SUMO-2			Polycomb proteins	SENP-5) Ulp2-like
SUMO-3			(PC-2, CBX-4)	(SENP6, SENP7)
			TOPORS KAP1	

pathway, SUMO proteases also regulate protein sumoylation (Fig. 1). Human cells contain a single SUMO E1 enzyme that functions as a heterodimer (SAE1, SAE2), as well as a single E2 enzyme (Ubc9). Several human SUMO E3 ligase.s have been reported (Table 1). However, unlike ubiquitin E3 enzymes, which can be recognized by HECT or RING motifs that are involved in ubiquitin transfer, common structural features required for SUMO E3 ligase activity are poorly understood, and the known SUMO E3 ligases vary significantly in both amino acid sequence and in proposed mechanisms of action. Unlike the HECT proteins that function as E3 enzymes in protein ubiquitination, none of the E3 enzymes identified in protein sumoylation function are capable of forming a thiolester with SUMO. In addition, the E2 SUMO enzyme binds to substrates directly and can promote protein sumoylation in vitro in the absence of an E3. While this finding initially suggested that E3 ligases might not be necessary for sumoylation, studies in yeast indicate a large dependency of cellular sumoylation on SUMO E3 ligases (Johnson and Gupta 2001).

Compared to ubiquitin E3 ligases, there are fewer known SUMO E3 ligases (Table 1). In addition, there appears to be overlap in substrate specificity for SUMO E3 ligases, evidenced by both the finding that cells lacking a particular E3 ligase are viable, and the overlap in substrates identified in proteomic studies (see below).

SUMO residues are removed from substrates by specific proteases (Fig. 1), which also function in removing short C-terminal sequences from newly synthesized SUMO paralogs, required to produce the C-terminal diglycine involved in the isopeptide bond with substrate lysines. SUMO proteases share a C-terminal domain that is distinct from ubiquitin proteases, but similar to a protease that removes the ubiquitin-like protein NEDD-8 (Gan-Erdene, Wilkinson 2003; Hay 2007). Several different SUMO proteases have been identified in humans, and can be grouped according to sequence homology with the two SUMO proteases in *S. cerevisiae*, Ulp1 and Ulp2 (Table 1) (Hay 2007). The six human SUMO proteases also exhibit distinct cellular localization as well as selectivity towards specific SUMO isoforms (Hay 2007). SENP3 and SENP5 localize in nucleoli and exhibit a preference for SUMO-2/3 (Di Bacco et al. 2006; Gong and Yeh 2006).

Cellular Roles for Sumoylation

Chromatin and Transcriptional Regulation

Chromatin remodeling and histone modifications have emerged as predominant mechanisms of gene expression control. Histones H3 and H4 contain specific lysine and serine residues that are covalently modified by phosphorylation, methylation, ubiquitination, and acetylation. The combinatorial nature of these histone modifica-tions reveal a "histone code" that serves as an epigenetic marking system that has an impact on chromatin-templated processes (Jenuwein and Allis 2001). For example, histone H4 is modified on Lys 5, Lys 8, Lys 12, Lys 16, and Lys 20. Among these five residues, four are regulated by acetylation, and have been shown to function in cell cycle progression, *Drosophila* embryogenesis, and transcriptional upregulation

(Clarke et al. 1999; Vogelauer et al. 2000; Ludlam et al. 2002). Modification of the amino-terminus of histone H3 can occur via methylation or phosphorylation. Phosphorylation of histone H3 Ser 10 inhibits Lys 9 methylation but is synergistically coupled with Lys 9 and Lys 14 acetylation during mitogenic and hormonal stimulation in mammalian cells (Cheung et al. 2000; Rea et al. 2000; Clayton et al. 2000; Lo et al. 2000; Jenuwein and Allis 2001). The phosphorylated-acetylated H3 denotes transcriptional activation in this modified state. Moreover, trimethylation of histone H3 Lys 4 by the Pc1/Pc2 polycomb complex results in transcriptional activation whereas dimethylation or trimethylation of histone H3 Lys 27 by the same complex results in transcriptional repression (Cao et al. 2002; Santos-Rosa et al. 2002). In addition to methylation, phosphorylation, and acetylation, SUMO modification of histone lysines has been recognized as an element of the histone code and chromatin organization (Shiio and Eisenman 2003; Iniguez-Lluhi 2006; Nathan et al. 2006). Shiio and Eisenman demonstrated that histone H4 is modified by SUMO *in vivo* and *in vitro*, with sumoylation mediated by Ubc9 (Shiio and Eisenman 2003). Furthermore, E3 SUMO ligases such as PIAS, Pc2, and TOPORS have been shown to be associated with chromatin (Hari et al. 2001; Capelson and Corces 2005; Pungaliya et al. 2007). Even though little is known regarding substrate specificity for E3 SUMO ligases, it has been suggested that they exhibit partial substrate specificity in part because E3 SUMO ligases may interact with specific pools of target substrates (Johnson 2004). Additionally, the local concentration of an E3 SUMO ligase may regulate substrate specificity (Reindle et al. 2006).

Genetic loss-of-function studies provide substantial evidence for SUMO isoforms in the regulation of higher order chromatin or chromosome structures. The *S. cerevisiae* ortholog of SUMO, SMT3 (Suppressor of Mif2 3 homolog 1), was first isolated as a high-copy suppressor of MIF2 mutations (Meluh and Koshland 1995). Studies have shown that loss of MIF2 function in yeast results in chromosome missegregation and mitotic delay (Brown et al. 1993), thus suggesting a role for SMT3 in centromere function. In vertebrate cells, SUMO-1 was identified as a suppressor of CENP-C which is the vertebrate homolog of Mif2 and a centromere protein that functions in kinetochore assembly (Fukagawa et al. 2001). Pmt3p, the *S. pombe* orthologue of SUMO-1, was shown to be required for maintenance of telomere length and chromosomal segregation (Tanaka et al. 1999). Furthermore, in yeast the evolutionarily conserved SUMO isopeptidase SMT4 has been shown to be essential for the regulation of chromosome condensation (Strunnikov et al. 2001). Moreover, this defect may be relieved by overexpression of Siz1, the yeast homologue of the mammalian PIAS proteins, suggesting that SMT4 and Siz1 are involved in a pathway of chromosome maintenance (Strunnikov et al. 2001). Finally, the gene for the suppressor of position-effect variegation, Su(var)2–10/Zimp, regulates chromosome structure and function and encodes the sole Drosophila homolog of the PIAS proteins (Hari et al. 2001).

Several transcriptional regulators are known to be sumoylated, typically resulting in inhibition of transcription. One mechanism by which sumoylation confers transcriptional repression is through the association of transcription factors with corepressors. For example, sumoylation of p300 has been shown to induce binding of HDAC6 (Girdwood et al. 2003), and sumoylation of histone H4 has been demonstrated to recruit HDAC1 (Shiio and Eisenman 2003). In Shiio and Eisenman's study, when Ubc9 was tethered to a promoter, transcriptional repression and decreased histone acetylation were observed, suggesting that sumoylation confers histone deacetylation and repression. Moreover, in another study, Yang et al. demonstrated that sumoylation of the transcription factor Elk1 recruits HDAC2 to Elk-1-regulated promoters, leading to transcriptional repression (Yang and Sharrocks 2004). Furthermore, repression mediated through SUMO conjugation to Elk-1 was alleviated by expression of the SUMO protease SSP3 (Yang et al. 2003) elucidating a mechanism for transcription control. Finally, reports have demonstrated that sumoylation of the corepressors HDAC1 and HDAC4 increases their deacetylase activity and transcriptional repression (David et al. 2002; Kirsh et al. 2002).

By contrast, sumoylation of transcriptional regulators may also result in activation of transcription. For instance, in prolonged hypoxia, cAMP-response element-binding protein (CREB) is posttranslationally modified by SUMO-1 (Comerford et al. 2003). Overexpression of SUMO stabilizes CREB in hypoxia and enhances CREB-dependent reporter gene activity (Comerford et al. 2003). Furthermore, sumoylation of the heat-shock transcription factors HSF1 and HSF2 results in enhanced DNA-binding activity and mutation of the acceptor lysine reduces the transcriptional activity of HSF1 (Goodson et al. 2001; Hong et al. 2001).

In cases of transcription factor regulation by sumoylation, there is precedence for a relatively small amount of sumoylation having a major impact on function, perhaps due to a cyclical rather than stoichiometric role for protein sumoylation (Fig. 2) (Johnson 2004; Hay 2005). For example, a recently manufactured transcription factor may be quickly sumoylated and incorporated into a repression complex in a SUMO-dependent manner (Girdwood et al. 2004). Then, SUMO-specific proteases remove the SUMO moiety, but the transcription factor remains "fixed" in the repression complex in a SUMO-independent manner (Girdwood et al. 2004). Eventually, a slow dissociation of the steady repression complex liberates the unmodified transcription factor. Therefore, SUMO may be vital for the initiation of repression, but not for preservation (Girdwood et al. 2004). Moreover, because sumoylation has also been implicated in protein–protein interactions, it is possible that SUMO conjugation of nuclear factors mediates protein complex formation (Johnson 2004). Therefore, as the model suggests, SUMO deconjugation would transpire after incorporation of the nuclear factor into the macromolecular complex (Hay 2005).

Regulation of Transcription Factors Implicated in Tumorigenesis

Several tumor suppressors and oncogenes such as promyelocytic leukemia protein (PML), p53, and Mdm2 have been associated with the sumoylation pathway. Mdm2, an oncogene that targets p53 for ubiquitination, is conjugated by SUMO-1 and this modification prevents the self-ubiquitination of Mdm2 (Buschmann et al. 2000). The tumor suppressor and transcription factor p53 has been well characterized as a sumoylated substrate with the SUMO-1 conjugation site mapping to Lys 386 (Muller et al. 2000). Multiple E3 ligases sumoylate p53, although only a small proportion of sumoylated p53 is typically present in cells (Kahyo et al. 2001;

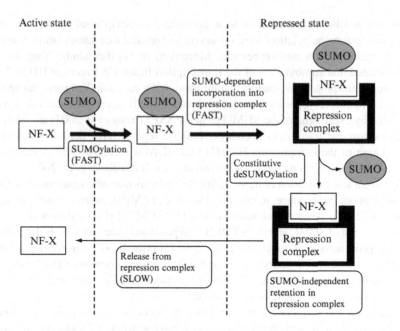

Fig. 2 Model depicting how a low fraction of SUMO conjugation can regulate the functional state of a protein. A recently synthesized nuclear transcription factor (NF-X) is quickly sumoylated and incorporated into a repression complex in a SUMO-dependent fashion. Then, SUMO-specific proteases remove the SUMO moiety, but the transcription factor remains retained in the repression complex in a SUMO-independent manner. Eventually, a slow dissociation of the steady repression complex liberates the unmodified transcription factor allowing the transcription factor to transfer back to the active state. Adapted with permission from (Girdwood et al. 2004).

Schmidt and Muller 2002; Weger et al. 2005). Interestingly, data regarding the functional consequence of p53 sumoylation by PIAS proteins points to both transcriptional activation and repression. Schmidt et al. demonstrated that PIAS proteins repress the transcriptional activity of p53 (Schmidt and Muller 2002). Moreover, PIASy was shown to repress p53-mediated activation of target genes (Nelson et al. 2001). On the other hand, through either the direct or tightly associated E3 ligase activity of PIAS1, sumoylation of p53 was shown to enhance its transcriptional activity (Gostissa et al. 1999; Rodriguez et al. 1999; Kahyo et al. 2001).

The tumor suppressor PML is also sumoylated. It has been suggested that SUMO modification of PML targets PML to the nuclear matrix and PML nuclear bodies (Muller et al. 1998; Zhong et al. 2000a). In acute promyelocytic leukemia (APL) cells, nuclear bodies are disrupted, which leads to aberrant localization of nuclear body proteins (Muller et al. 1998). In the majority of acute promyelocytic leukemia cases, a chromosomal translocation occurs that involves the retinoic acid receptor alpha (RARα) gene and the PML gene (Piazza et al. 2001). The treatment of APL patients with all-*trans* retinoic acid results in clinical remission associated with blast cell differentiation and reformation of the PML nuclear bodies (Piazza et al. 2001). Studies done in primary *PML*$^{-/-}$ cells show that in the absence of PML, several nuclear body proteins such as Sp100, CBP, ISG20, Daxx, and SUMO fail to accumulate in the nuclear body and acquire aberrant localization patterns (Zhong et al. 2000a). In addition, a PML mutant that can no longer be modified by SUMO displays an irregular nuclear localization pattern (Zhong et al. 2000). These data suggest that PML is required for the proper formation of the nuclear body and that conjugation to SUMO is required for PML function.

Three different roles for the PML nuclear body have been proposed in the control of transcription: titration, modification, and compartmentalization (Zhong et al. 2000b). In titration, the nuclear body may regulate the concentration of nuclear transcription factors by sequestration, while in modification sumoylation or acetylation may occur in the nuclear body, thereby regulating their transcriptional activity. In compartmentalization, the nuclear bodies represent centers where dual-function factors can exert either one of their functions. For example, RAR/RXR actively repress transcription through interaction with corepressors and histone deacetylases (Grunstein 1997; Rietveld et al. 2001; Urnov et al. 2001), whereas when bound to retinoic acid, they can activate transcription through interaction with coactivators such as CBP (Kamei et al. 1996), which is found in the PML nuclear body. These three proposed models imply that the PML nuclear body regulates the transcriptional activity of particular transcription factors and their target genes.

PML nuclear bodies have also been implicated as sites of assembly of transcriptionally repressive structures that involve sumoylation (Hay 2005). For example, sumoylation of HIPK2 is associated with localization of the protein in PML nuclear bodies, as well as transcriptional repression (Kim et al. 1999). PML nuclear bodies have also been implicated as sites of heterochromatin remodeling at G2 phase in lymphocytes of patients with immunodeficiency, centromeric instability and facial dysmorphy (ICF) syndrome (Luciani et al. 2006). There is also evidence that RNA polymerase II and its nascent transcripts localize in PML nuclear bodies, suggesting that active transcription takes place in these structures (LaMorte et al. 1998; von Mikecz et al. 2000). Additionally, the corepressor mSin3A was shown to interact with PML and this interaction was proposed to mediate the transcriptional repression by tumor suppressor Mad (Khan et al. 2001). Moreover, mSin3A has been shown to be sumoylated by the putative tumor suppressor and E3 SUMO ligase TOPORS which also localizes to PML nuclear bodies (Rasheed et al. 2002; Saleem et al. 2004; Pungaliya et al. 2007). Therefore, TOPORS may regulate substrate function by sumoylation-dependent recruitment to PML nuclear bodies.

Role of Sumoylation in Eukaryotes

Loss-of-function studies in model organisms (e.g., mice) are particularly helpful when considering pathway inhibition as a therapeutic strategy. Several of these kinds of studies have been reported for components of the sumoylation pathway. In general, these studies indicate that sumoylation is essential for viability, and argue that sumoylation is particularly important in chromatin regulation and cellular sensitivity to DNA damage.

For example, in fission yeast, deficiency of the SUMO E2 Hus5/Ubc9 results in abnormal chromosome segregation (al-Khodairy et al. 1995), and in loss of gene silencing as well as altered histone modification patterns at heterochromatic regions (Shin et al. 2005). In mice, loss of Ubc9 results in embryonic lethality. Murine blasto-cyst cells lacking Ubc9 exhibit mitotic chromosome defects, as well as disruption of nucleoli and promyelocytic leukemia nuclear bodies (Nacerddine et al. 2005).

In contrast to the SUMO E2, studies of SUMO E3 ligases in mice indicate that loss of a specific E3 enzyme does not result in embryonic lethality, presumably related to a redundancy among SUMO E3 ligases. However, similar to phenotypes associated with loss of Ubc9, chromosomal instability and alterations in chromatin have been reported in cells deficient in SUMO E3 ligases. Mice lacking the SUMO E3 ligase PIASy appear phenotypically normal (Wong et al. 2004), while mice lacking PIAS1 exhibit an increase in perinatal lethality and reduced weight (Liu et al. 2004). Knockdown of the SUMO E3 ligase PIASγ in human HeLa cancer cells results in a defect in mitotic chromosome segregation, possibly due to loss of topoisomerase II sumoylation (Diaz-Martinez et al. 2006). Complete deficiency of the dual SUMO and ubiquitin E3 ligase *Topors* results in perinatal mortality in mice, with mice heterozygus for a mutant *Topors* allele exhibiting an increased rate of malignancy (Marshall, H., Rubin, E., et al., submitted). In addition, loss of *Topors* in murine embryonic fibroblasts results in a defect in mitotic chromosome segregation associated with an alteration in the localization of HP1α to pericentric heterochromatin (Marshall, H., Rubin, E., et al., submitted). Loss of SUMO E3 ligases is also associated with resistance to anticancer drugs. Yeast cells deficient in Siz 1 are resistant to the topoisomerase II-targeting drug doxorubicin (Huang et al. 2007), and murine cells deficient in *TOPORS* are resistant to the histone deacetylase inhibitor trichostatin A (Marshall, H. et al., submitted).

Loss of the SUMO protease Ulp1 in budding yeast is lethal (Li and Hochstrasser 1999), whereas loss of the orthologous protease in fission yeast confers sensitivity to ultraviolet-light-induced DNA damage (Taylor et al. 2002). Yeast cells lacking Ulp2 are viable but exhibit chromosomal instability, and Ulp2 has a role in mitotic chromo-somal segregation (Bachant et al. 2002; Stead et al. 2003), as well as in proper mitotic spindle formation following DNA-damage-induced cell cycle arrest (Schwartz et al. 2007). Loss of function studies also implicate a SUMO protease (SENP1) in angio-genesis: loss of SENP1 results in embryonic lethality due to severe anemia, related to an increase the sumoylation of HIF1α, which results in instability of this transcription factor and loss of transcription of HIF1α-dependent genes (Cheng et al. 2007).

With regard to SUMO orthologs, mice with a decrease in SUMO-1 due to the presence of a hypomorphic allele exhibit an increase in perinatal mortality (Alkuraya et al. 2006). In addition, heterozygotes were found to have defects in lip and palate development, and in humans haploinsufficiency of SUMO-1 is implicated in cleft lip and palate (Alkuraya et al. 2006). These results indicate a nonredundant role for the SUMO-1 isoform in mammals. Although knockouts of SUMO-2 or SUMO-3 in mice have not been reported, specific conjugation of topoisomerase II by SUMO-2/3, as opposed SUMO-1, has been identified, supporting the concept of unique roles for different SUMO isoforms (Azuma et al. 2003). In addition, in contrast to SUMO-1, cells contain free SUMO-2/3 that can be used for protein sumoylation upon stress, such as heat shock (Saitoh and Hinchey 2000).

Additional information regarding specific cellular functions for different SUMO isoforms relates to polymeric chain formation. Although polymeric SUMO-1 chains have been described (Pichler et al. 2002; Hammer et al. 2007), unlike SUMO-2 and SUMO-3, SUMO-1 lacks an internal ΨKXE motif, and is less efficient in chain formation (Tatham et al. 2001). Polymeric SUMO-2/3 chains have been detected in cells and are implicated in targeting proteins for ubiquitination and subsequent proteasome-mediated degradation of a particular sumoylated protein (Ii et al. 2007; Prudden et al. 2007; Sun et al. 2007; Uzunova et al. 2007; Xie et al. 2007). Notably, yeast cells that express only a mutant form of SUMO, in which lysines implicated in polymeric chain formation were mutated, are viable, suggesting that polymeric SUMO chain formation is not essential, at least in *S. cerevisiae* (Bylebyl et al. 2003). However, *S. pombe* cells that accumulate SUMO chains as a result of loss of a specific ubiquitin ligase (Slx8-Rfp) exhibit genomic instability and sensitivity to DNA-damaging agents (Prudden et al. 2007). Similarly, the human RNF4 protein, which is involved in the response to DNA damage and is the human ortholog of this SUMO-chain ubiquitin ligase, is required for degradation of polysumoylated PML (Tatham et al. 2008).

In addition to the loss-of-function studies described above, several groups have used proteomic approaches to identify sumoylated proteins, thereby gaining insight into the cellular role of SUMO isoforms. However, it should be noted that these approaches are complicated by the typical low percentage of sumoylation of a given substrate that is found in proliferating cells. The apparent discrepancy between the low percentage of steady-state sumoylation versus the importance of sumoylation in protein function may be explained by a role for sumoylation in state switching, e.g., cellular localization, that does not require persistence of SUMO conjugation (Hay 2005). Proteomic studies implicate sumoylation in a broad spectrum of cellular processes, including protein degradation, transcriptional regulation, heterochromatin formation, and DNA repair (Li et al. 2004; Vertegaal et al. 2004; Gocke et al. 2005; Rosas-Acosta et al. 2005).

Additional information is available regarding the role of sumoylation in the biology of specific substrates. In the case of IκBα, these studies indicate that monomeric SUMO conjugation can block protein ubiquitination by competition with ubiquitin for the same lysine residue (Desterro et al. 1998). Thus, one clear role for sumoylation is to protect proteins from ubiquitin-mediated proteasomal degradation. Other studies of specific sumoylation substrates indicate that sumoylation may regulate protein–protein interactions, subcellular localization, or DNA binding by transcription factors (Hay 2005; Chou et al. 2007).

Alteration of SUMO Pathways in Human Diseases

Posttranslational modification by ubiquitin or ubiquitin-like modifiers such as SUMO may regulate a protein's function or subcellular localization. Under normal conditions, these modifications help to maintain homeostasis and promote cell survival.

However, dysregulation of protein sumoylation has been linked to several human diseases (Kerscher et al. 2006). In particular, recent studies have demonstrated that components of the sumoylation pathway may be involved in the pathogenesis of neurodegenerative disorders (Martin et al. 2007), clefts of the lip and/or palate (Pauws and Stanier 2007), retinitis pigmentosa (Chakarova et al. 2007), and cancer (Seeler et al. 2007).

Several neurological disorders, including Alzheimer's disease and Parkinson's disease, are characterized by the accumulation of ubiquitinated protein aggregates within neurons (McNaught et al. 2001; Upadhya and Hegde 2007). However, it is not clear why these ubiquitinated proteins are not degraded. Interestingly, several studies have implicated altered SUMO modification in neurodegenerative diseases, including multiple system atrophy (Pountney et al. 2005), neuronal intranuclear inclusion disease (NIID) (Pountney et al. 2003), and polyglutamine diseases such as Huntington's disease (HD) (Steffan et al. 2004) and spinocerebellar ataxia type 3 (Ueda et al. 2002). In a *Drosophila* model of polyglutamine disease, overexpression of a mutant form of the SUMO-activating enzyme (Uba2) caused enhanced eye degeneration, suggesting that disruption of SUMO pathway may increase polyglutamine toxicity (Chan et al. 2002). Pountney et al. reported SUMO-1 immunostaining in brain sections of NIID, HD, and spinocerebellar ataxia type 3 (Pountney et al. 2003).

Studies of Huntingtin, the protein implicated in Huntington's disease, also implicate the SUMO-1 pathway in the cellular response to aggregated proteins. Because sumoylation can increase Huntingtin accumulation, toxic oligomer concentrations may develop (Steffan et al. 2004). Robust punctuate SUMO-1 immunostaining has been shown to mark subdomains of the oligodendroglial intracytoplasmic protein aggregates in the α-synucleinopathy disease, multiple system aptrophy (Pountney et al. 2005). Moreover, cell culture experiments indicated that nuclear and peri-nuclear accumulation of SUMO-1 aggregates could be induced in glioma cells by chemical inhibition of proteasomal protein degradation, indicating that accumulation of SUMO-1 modified proteins in the nucleus may be linked to dysfunction of the proteasome machinery (Pountney et al. 2005).

In Parkinson's disease, Parkin and DJ-1 are closely linked to the sumoylation pathway. Parkin, which is an E3 ubiquitin ligase, interacts noncovalently with SUMO-1 (Um and Chung 2006). Moreover, Parkin regulates the turnover of the SUMO E3 ligase RanBP2 in a proteasome-dependent manner (Um et al. 2006). DJ-1 is an oxidative stress sensor, with loss of function of this protein implicated in early onset Parkinson's disease (Jenner 2003; Shinbo et al. 2006). Recent studies have shown that DJ-1 is sumoylated on K130 by SUMO E3 ligases PIASxα and PIASy (Shinbo et al. 2006). Furthermore, mutation of K130 abrogated the function of DJ-1 in *ras*-dependent transformation, cell growth initiation, and anti-UV-induced apoptosis, indicating that proper SUMO-1 conjugation to DJ-1 is essential for DJ-1 to exert its full activities (Shinbo et al. 2006).

Finally, SUMO-3 overexpression has been shown to reduce the production of neurotoxin amyloid β peptide, a peptide generated from amyloid precursor protein, which is central to Alzheimer's disease (Li et al. 2003). Poly-SUMO-3 chains were shown to be important regulators of the proteolytic processing of amyloid precursor protein because overexpression of a mutant SUMO-3 incapable of forming polyconjugates increased amyloid β peptide in cells (Li et al. 2003).

Additional diseases in which alterations in the sumoylation pathway are implicated include clefts of the lip and/or palate (Pauws and Stanier 2007), and retinitis pigmentosa (Chakarova et al. 2007). In a patient with cleft lip and palate, a translocation breakpoint in the SUMO-1 gene was identified (Alkuraya et al. 2006). Also, transcription factor TBX22, which is essential for normal craniofacial development, requires sumoylation for its transcriptional repressor function (Andreou et al. 2007). Interestingly, TBX22 missense mutations found in patients with X-linked cleft palate render the protein unable to be sumoylated (Andreou et al. 2007). TOPORS or topoisomerase I-binding arginine-serine rich protein, (Haluska et al. 1999; Rasheed et al. 2002) is the first example of a protein with both ubiquitin and SUMO-1 E3 ligase activity (Weger et al. 2003; Rajendra et al. 2004; Weger et al. 2005). In addition to being implicated as a tumor suppressor in several different malignancies (Oyanagi et al. 2004; Saleem et al. 2004; Bredel et al. 2005; Shinbo et al. 2005), mutations in the *TOPORS* gene have been reported in patients with autosomal dominant retinitis pigmentosa (Chakarova et al. 2007).

Finally, given the roles of the sumoylation pathway in chromatin regulation and in sensitivity to DNA damage, it is not surprising that alterations in the sumoylation pathway have been implicated in the etiology of several different kinds of cancer (Seeler et al. 2007), as well as in cancer metastasis (Kim and Baek 2006).

Targeting the Sumoylation Pathway in Human Diseases

Although there are not yet examples of specific SUMO-targeting therapeutics that are in the clinic, frequent involvement of sumoylation in various human diseases has engendered interest in targeting various components of the sumoylation pathway, particularly in neurodegenerative diseases and cancer. Since the sumoylation pathway consists of several enzymatic steps, reducing the expression of SUMO precursors, inhibiting Ubc9 or E3 SUMO ligases, or enhancing isopeptidase activity are all potential therapeutic strategies.

With regard to neurodegenerative diseases, the involvement of SUMO-3 in Alzheimer's disease suggests that targeting this specific SUMO isotype may be beneficial in this disease (Li et al. 2003).

With regard to cancer, proteins with increased expression in specific cancers may be important in the growth and progression of these cancers and are thus attractive therapeutic targets. There are now several examples of successful development of drugs designed to specifically target overexpressed proteins (e.g., trastuzumab, which targets HER2, for breast cancer, and imatinib, which targets c-kit, for gastrointestinal stromal tumors). As described above, loss of specific sumoylation pathway components is implicated in genomic instability that can lead to malignancy. However, as discussed below, there are examples of overexpression of specific components of the sumoylation pathway in certain human cancers, suggesting that these could be therapeutic targets.

In addition, the success of histone deacetylase inhibitors in the treatment of cancer provides a precedent for targeting proteins involved in chromatin regulation, including components of the sumoylation pathway (Marks et al. 2001; Drummond et al. 2005).

Ubc9 has been proposed as a therapeutic target. Overexpression of Ubc9 in breast cancer xenografts increases tumor growth, whereas overexpression of a dominant negative mutant Ubc9 (containing a substitution of serine for cysteine 93) results in decreased tumor growth (Mo et al. 2005). In addition, Ubc9 transcript expression was found to be increased in ovarian cancer tissues compared to adjacent normal tissue (Mo et al. 2005). Although increased Ubc9 activity would be expected to alter the function or location of many proteins, increased expression of the oncogene *BCL-2* was found in cells overexpressing Ubc9 (Mo et al. 2005). These findings suggest that increased activity of Ubc9 is selected for during carcinogenesis.

Certain E3 SUMO ligases are also overexpressed in cancer and may be therapeutic targets. For example, PIAS3 expression is increased in several cancers, including breast and prostate cancer (Wang and Banerjee 2004).

SUMO proteases have also been implicated as therapeutic targets. SENP1 was identified as involved in a translocation of chromosomes 12 and 15 in a patient with a teratoma, resulting in a fusion protein with the mesoderm development gene *MESDC2* (Veltman et al. 2005), and was also identified in a screen for genes with increased expression in thyroid oncocytic tumors relative to normal thyroid tissue (Jacques et al. 2005). Increased expression of SENP1 RNA was also observed in some (60%: 29/43 cases) of prostate cancer tissues relative to normal prostate tissue (Cheng et al. 2006). In addition, SENP1 was shown to be required for the stability of HIF1α during hypoxia (Cheng et al. 2007). Mice lacking SENP1 die during embryogenesis as a result of severe anemia, related to a decrease in HIF1α protein stability during hypoxia (Cheng et al. 2007). Additional experiments suggest that as discussed above for polymeric SUMO chains, sumoylation targets HIF1α for proteasome-dependent degradation (Cheng et al. 2007). Thus, targeting SENP1 can be added to list of anti-HIF1α, antiangiogenic efforts in cancer therapy (Semenza 2007).

In summary, the finding that alterations in the sumoylation pathway are important in several human diseases makes this pathway an attractive target for therapeutics. Nevertheless, the vital role of Ubc9 in various cellular functions suggests that targeting the SUMO E1 or E2 enzymes is likely to be associated with significant toxicity. By contrast, continued advances in understanding the specificity of E3 SUMO ligases and SUMO proteases, as well as the effects of sumoylation on the function of specific sumoylated proteins, may allow development of small molecules that correct sumoylation defects in diseased cells, while having minimal effects on normal cells.

References

al-Khodairy, F., T. Enoch, et al. (1995). "The Schizosaccharomyces pombe hus5 gene encodes a ubiquitin conjugating enzyme required for normal mitosis." *J Cell Sci* **108** (Pt 2): 475–86.

Alkuraya, F. S., I. Saadi, et al. (2006). "SUMO1 haploinsufficiency leads to cleft lip and palate." *Science* **313**(5794): 1751.

Andreou, A. M., E. Pauws, et al. (2007). "TBX22 missense mutations found in patients with X-linked cleft palate affect DNA binding, sumoylation, and transcriptional repression." *Am J Hum Genet* **81**(4): 700–12.

Azuma, Y., A. Arnaoutov, et al. (2003). "SUMO-2/3 regulates topoisomerase II in mitosis." *J Cell Biol* **163**(3): 477–87.

Bachant, J., A. Alcasabas, et al. (2002). "The SUMO-1 isopeptidase Smt4 is linked to centromeric cohesion through SUMO-1 modification of DNA topoisomerase II." *Mol Cell* **9**(6): 1169–82.

Bayer, P., A. Arndt, et al. (1998). "Structure determination of the small ubiquitin-related modifier SUMO-1." *J Mol Biol* **280**(2): 275–86.

Bredel, M., C. Bredel, et al. (2005). "High-resolution genome-wide mapping of genetic alterations in human glial brain tumors." *Cancer Res* **65**(10): 4088–96.

Brown, M. T., L. Goetsch, et al. (1993). "MIF2 is required for mitotic spindle integrity during anaphase spindle elongation in Saccharomyces cerevisiae." *J Cell Biol* **123**(2): 387–403.

Buschmann, T., S. Y. Fuchs, et al. (2000). "SUMO-1 modification of Mdm2 prevents its self-ubiquitination and increases Mdm2 ability to ubiquitinate p53." *Cell* **101**(7): 753–62.

Bylebyl, G. R., I. Belichenko, et al. (2003). "The SUMO isopeptidase Ulp2 prevents accumulation of SUMO chains in yeast." *J Biol Chem* **278**(45): 44113–20.

Cao, R., L. Wang, et al. (2002). "Role of histone H3 lysine 27 methylation in Polycomb-group silencing." *Science* **298**(5595): 1039–43.

Capelson, M. and V. G. Corces (2005). "The ubiquitin ligase dTopors directs the nuclear organization of a chromatin insulator." *Mol Cell* **20**(1): 105–16.

Chakarova, C. F., M. G. Papaioannou, et al. (2007). "Mutations in TOPORS cause autosomal dominant retinitis pigmentosa with perivascular retinal pigment epithelium atrophy." *Am J Hum Genet* **81**(5): 1098–103.

Chan, H. Y., J. M. Warrick, et al. (2002). "Genetic modulation of polyglutamine toxicity by protein conjugation pathways in Drosophila." *Hum Mol Genet* **11**(23): 2895–904.

Cheng, J., T. Bawa, et al. (2006). "Role of desumoylation in the development of prostate cancer." *Neoplasia* **8**(8): 667–76.

Cheng, J., X. Kang, et al. (2007). "SUMO-specific protease 1 is essential for stabilization of HIF1alpha during hypoxia." *Cell* **131**(3): 584–95.

Cheung, P., K. G. Tanner, et al. (2000). "Synergistic coupling of histone H3 phosphorylation and acetylation in response to epidermal growth factor stimulation." *Mol Cell* **5**(6): 905–15.

Chou, C. C., C. Chang, et al. (2007). "Small ubiquitin-like modifier modification regulates the DNA binding activity of glial cell missing Drosophila homolog a." *J Biol Chem* **282**(37): 27239–49.

Clarke, A. S., J. E. Lowell, et al. (1999). "Esa1p is an essential histone acetyltransferase required for cell cycle progression." *Mol Cell Biol* **19**(4): 2515–26.

Clayton, A. L., S. Rose, et al. (2000). "Phosphoacetylation of histone H3 on c-fos- and c-jun-associated nucleosomes upon gene activation." *Embo J* **19**(14): 3714–26.

Comerford, K. M., M. O. Leonard, et al. (2003). "Small ubiquitin-related modifier-1 modification mediates resolution of CREB-dependent responses to hypoxia." *Proc Natl Acad Sci U S A* **100**(3): 986–91.

David, G., M. A. Neptune, et al. (2002). "SUMO-1 modification of histone deacetylase 1 (HDAC1) modulates its biological activities." *J Biol Chem* **277**(26): 23658–63.

Desterro, J. M., M. S. Rodriguez, et al. (1998). "SUMO-1 modification of IkappaBalpha inhibits NF-kappaB activation." *Mol Cell* **2**(2): 233–9.

Di Bacco, A., J. Ouyang, et al. (2006). "The SUMO-specific protease SENP5 is required for cell division." *Mol Cell Biol* **26**(12): 4489–98.

Diaz-Martinez, L. A., J. F. Gimenez-Abian, et al. (2006). "PIASgamma is required for faithful chromosome segregation in human cells." *PLoS ONE* **1**: e53.

Drummond, D. C., C. O. Noble, et al. (2005). "Clinical development of histone deacetylase inhibitors as anticancer agents." *Annu Rev Pharmacol Toxicol* **45**: 495–528.

Fukagawa, T., V. Regnier, et al. (2001). "Creation and characterization of temperature-sensitive CENP-C mutants in vertebrate cells." *Nucleic Acids Res* **29**(18): 3796–803.

Girdwood, D., D. Bumpass, et al. (2003). "P300 transcriptional repression is mediated by SUMO modification." *Mol Cell* **11**(4): 1043–54.

Girdwood, D. W., M. H. Tatham, et al. (2004). "SUMO and transcriptional regulation." *Semin Cell Dev Biol* **15**(2): 201–10.

Gocke, C. B., H. Yu, et al. (2005). "Systematic identification and analysis of mammalian small ubiquitin-like modifier substrates." *J Biol Chem* **280**(6): 5004–12.

Gong, L. and E. T. Yeh (2006). "Characterization of a family of nucleolar SUMO-specific proteases with preference for SUMO-2 or SUMO-3." *J Biol Chem* **281**(23): 15869–77.

Goodson, M. L., Y. Hong, et al. (2001). "Sumo-1 modification regulates the DNA binding activity of heat shock transcription factor 2, a promyelocytic leukemia nuclear body associated transcription factor." *J Biol Chem* **276**(21): 18513–8.

Gostissa, M., A. Hengstermann, et al. (1999). "Activation of p53 by conjugation to the ubiquitin-like protein SUMO-1." *Embo J* **18**(22): 6462–71.

Grunstein, M. (1997). "Histone acetylation in chromatin structure and transcription." *Nature* **389**(6649): 349–52.

Haluska, P., Jr., A. Saleem, et al. (1999). "Interaction between human topoisomerase I and a novel RING finger/arginine-serine protein." *Nucleic Acids Res* **27**(12): 2538–44.

Hammer, E., R. Heilbronn, et al. (2007). "The E3 ligase Topors induces the accumulation of polysumoylated forms of DNA topoisomerase I in vitro and in vivo." *FEBS Lett* **581**(28): 5418–24.

Hari, K. L., K. R. Cook, et al. (2001). "The Drosophila Su(var)2-10 locus regulates chromosome structure and function and encodes a member of the PIAS protein family." *Genes Dev* **15**(11): 1334–48.

Hay, R. T. (2005). "SUMO: a history of modification." *Mol Cell* **18**(1): 1–12.

Hay, R. T. (2007). "SUMO-specific proteases: a twist in the tail." *Trends Cell Biol* **17**(8): 370–6.

Hecker, C. M., M. Rabiller, et al. (2006). "Specification of SUMO1 and SUMO2 interacting motifs." *J Biol Chem* **281**(23): 16117–27.

Hong, Y., R. Rogers, et al. (2001). "Regulation of heat shock transcription factor 1 by stress-induced SUMO-1 modification." *J Biol Chem* **276**(43): 40263–7.

Huang, R. Y., D. Kowalski, et al. (2007). "Small ubiquitin-related modifier pathway is a major determinant of doxorubicin cytotoxicity in Saccharomyces cerevisiae." *Cancer Res* **67**(2): 765–72.

Ii, T., J. Fung, et al. (2007). "The yeast Slx5-Slx8 DNA integrity complex displays ubiquitin ligase activity." *Cell Cycle* **6**(22): 2800–9.

Iniguez-Lluhi, J. A. (2006). "For a healthy histone code, a little SUMO in the tail keeps the acetyl away." *ACS Chem Biol* **1**(4): 204–6.

Jacques, C., O. Baris, et al. (2005). "Two-step differential expression analysis reveals a new set of genes involved in thyroid oncocytic tumors." *J Clin Endocrinol Metab* **90**(4): 2314–20.

Jenner, P. (2003). "Oxidative stress in Parkinson's disease." *Ann Neurol* **53 Suppl 3**: S26–36; discussion S36–8.

Jenuwein, T. and C. D. Allis (2001). "Translating the histone code." *Science* **293**(5532): 1074–80.

Johnson, E. S. (2004). "Protein modification by SUMO." *Annu Rev Biochem* **73**: 355–82.

Johnson, E. S. and A. A. Gupta (2001). "An E3-like factor that promotes SUMO conjugation to the yeast septins." *Cell* **106**(6): 735–44.

Kahyo, T., T. Nishida, et al. (2001). "Involvement of PIAS1 in the sumoylation of tumor suppressor p53." *Mol Cell* **8**(3): 713–8.

Kamei, Y., L. Xu, et al. (1996). "A CBP integrator complex mediates transcriptional activation and AP-1 inhibition by nuclear receptors." *Cell* **85**(3): 403–14.

Kerscher, O., R. Felberbaum, et al. (2006). "Modification of proteins by ubiquitin and ubiquitin-like proteins." *Annu Rev Cell Dev Biol* **22**: 159–80.

Khan, M. M., T. Nomura, et al. (2001). "Role of PML and PML-RARalpha in Mad-mediated transcriptional repression." *Mol Cell* **7**(6): 1233–43.

Kim, K. I. and S. H. Baek (2006). "SUMOylation code in cancer development and metastasis." *Mol Cells* **22**(3): 247–53.

Kim, Y. H., C. Y. Choi, et al. (1999). "Covalent modification of the homeodomain-interacting protein kinase 2 (HIPK2) by the ubiquitin-like protein SUMO-1." *Proc Natl Acad Sci U S A* **96**(22): 12350–5.

Kirsh, O., J. S. Seeler, et al. (2002). "The SUMO E3 ligase RanBP2 promotes modification of the HDAC4 deacetylase." *Embo J* **21**(11): 2682–91.

LaMorte, V. J., J. A. Dyck, et al. (1998). "Localization of nascent RNA and CREB binding protein with the PML-containing nuclear body." *Proc Natl Acad Sci U S A* **95**(9): 4991–6.

Li, S. J. and M. Hochstrasser (1999). "A new protease required for cell-cycle progression in yeast." *Nature* **398**(6724): 246–51.

Li, T., E. Evdokimov, et al. (2004). "Sumoylation of heterogeneous nuclear ribonucleoproteins, zinc finger proteins, and nuclear pore complex proteins: a proteomic analysis." *Proc Natl Acad Sci U S A* **101**(23): 8551–6.

Li, Y., H. Wang, et al. (2003). "Positive and negative regulation of APP amyloidogenesis by sumoylation." *Proc Natl Acad Sci U S A* **100**(1): 259–64.

Liu, B., S. Mink, et al. (2004). "PIAS1 selectively inhibits interferon-inducible genes and is important in innate immunity." *Nat Immunol* **5**(9): 891–8.

Lo, W. S., R. C. Trievel, et al. (2000). "Phosphorylation of serine 10 in histone H3 is functionally linked in vitro and in vivo to Gcn5-mediated acetylation at lysine 14." *Mol Cell* **5**(6): 917–26.

Luciani, J. J., D. Depetris, et al. (2006). "PML nuclear bodies are highly organised DNA-protein structures with a function in heterochromatin remodelling at the G2 phase." *J Cell Sci* **119**(Pt 12): 2518–31.

Ludlam, W. H., M. H. Taylor, et al. (2002). "The acetyltransferase activity of CBP is required for wingless activation and H4 acetylation in *Drosophila melanogaster*." *Mol Cell Biol* **22**(11): 3832–41.

Marks, P., R. A. Rifkind, et al. (2001). "Histone deacetylases and cancer: causes and therapies." *Nat Rev Cancer* **1**(3): 194–202.

Martin, S., K. A. Wilkinson, et al. (2007). "Emerging extranuclear roles of protein SUMOylation in neuronal function and dysfunction." *Nat Rev Neurosci* **8**(12): 948–59.

McNaught, K. S., C. W. Olanow, et al. (2001). "Failure of the ubiquitin-proteasome system in Parkinson's disease." *Nat Rev Neurosci* **2**(8): 589–94.

Meluh, P. B. and D. Koshland (1995). "Evidence that the MIF2 gene of *Saccharomyces cerevisiae* encodes a centromere protein with homology to the mammalian centromere protein CENP-C." *Mol Biol Cell* **6**(7): 793–807.

Mo, Y. Y., Y. Yu, et al. (2005). "A role for Ubc9 in tumorigenesis." *Oncogene* **24**(16): 2677–83.

Muller, S., M. J. Matunis, et al. (1998). "Conjugation with the ubiquitin-related modifier SUMO-1 regulates the partitioning of PML within the nucleus." *Embo J* **17**(1): 61–70.

Muller, S., M. Berger, et al. (2000). "c-Jun and p53 activity is modulated by SUMO-1 modification." *J Biol Chem* **275**(18): 13321–9.

Nacerddine, K., F. Lehembre, et al. (2005). "The SUMO pathway is essential for nuclear integrity and chromosome segregation in mice." *Dev Cell* **9**(6): 769–79.

Nathan, D., K. Ingvarsdottir, et al. (2006). "Histone sumoylation is a negative regulator in Saccharomyces cerevisiae and shows dynamic interplay with positive-acting histone modifications." *Genes Dev* **20**(8): 966–76.

Nelson, V., G. E. Davis, et al. (2001). "A putative protein inhibitor of activated STAT (PIASy) interacts with p53 and inhibits p53-mediated transactivation but not apoptosis." *Apoptosis* **6**(3): 221–34.

Oyanagi, H., K. Takenaka, et al. (2004). "Expression of LUN gene that encodes a novel RING finger protein is correlated with development and progression of non-small cell lung cancer." *Lung Cancer* **46**(1): 21–8.

Pauws, E. and P. Stanier (2007). "FGF signalling and SUMO modification: new players in the aetiology of cleft lip and/or palate." *Trends Genet* **23**(12): 631–40.

Piazza, F., C. Gurrieri, et al. (2001). "The theory of APL." *Oncogene* **20**(49): 7216–22.

Pichler, A., A. Gast, et al. (2002). "The nucleoporin RanBP2 has SUMO1 E3 ligase activity." *Cell* **108**(1): 109–20.

Pountney, D. L., F. Chegini, et al. (2005). "SUMO-1 marks subdomains within glial cytoplasmic inclusions of multiple system atrophy." *Neurosci Lett* **381**(1–2): 74–9.

Pountney, D. L., Y. Huang, et al. (2003). "SUMO-1 marks the nuclear inclusions in familial neuronal intranuclear inclusion disease." *Exp Neurol* **184**(1): 436–46.

Prudden, J., S. Pebernard, et al. (2007). "SUMO-targeted ubiquitin ligases in genome stability." *Embo J* **26**(18): 4089–101.

Pungaliya, P., D. Kulkarni, et al. (2007). "TOPORS functions as a SUMO-1 E3 ligase for chromatin-modifying proteins." *J Proteome Res* **6**(10): 3918–23.

Rajendra, R., D. Malegaonkar, et al. (2004). "Topors functions as an E3 ubiquitin ligase with specific E2 enzymes and ubiquitinates p53." *J Biol Chem* **279**(35): 36440–4.

Rasheed, Z. A., A. Saleem, et al. (2002). "The topoisomerase I-binding RING protein, topors, is associated with promyelocytic leukemia nuclear bodies." *Exp Cell Res* **277**(2): 152–60.

Rea, S., F. Eisenhaber, et al. (2000). "Regulation of chromatin structure by site-specific histone H3 methyltransferases." *Nature* **406**(6796): 593–9.

Reindle, A., I. Belichenko, et al. (2006). "Multiple domains in Siz SUMO ligases contribute to substrate selectivity." *J Cell Sci* **119**(Pt 22): 4749–57.

Rietveld, L. E., E. Caldenhoven, et al. (2001). "Avian erythroleukemia: a model for corepressor function in cancer." *Oncogene* **20**(24): 3100–9.

Rodriguez, M. S., C. Dargemont, et al. (2001). "SUMO-1 conjugation in vivo requires both a consensus modification motif and nuclear targeting." *J Biol Chem* **276**(16): 12654–9.

Rodriguez, M. S., J. M. Desterro, et al. (1999). "SUMO-1 modification activates the transcriptional response of p53." *Embo J* **18**(22): 6455–61.

Rosas-Acosta, G., W. K. Russell, et al. (2005). "A universal strategy for proteomic studies of SUMO and other ubiquitin-like modifiers." *Mol Cell Proteomics* **4**(1): 56–72.

Saitoh, H. and J. Hinchey (2000). "Functional heterogeneity of small ubiquitin-related protein modifiers SUMO-1 versus SUMO-2/3." *J Biol Chem* **275**(9): 6252–8.

Saleem, A., J. Dutta, et al. (2004). "The topoisomerase I- and p53-binding protein topors is differentially expressed in normal and malignant human tissues and may function as a tumor suppressor." *Oncogene* **23**(31): 5293–300.

Sampson, D. A., M. Wang, et al. (2001). "The SUMO-1 consensus sequence mediates Ubc9 binding and is essential for SUMO-1 modification." *J Biol Chem* **19**: 19.

Santos-Rosa, H., R. Schneider, et al. (2002). "Active genes are tri-methylated at K4 of histone H3." *Nature* **419**(6905): 407–11.

Schmidt, D. and S. Muller (2002). "Members of the PIAS family act as SUMO ligases for c-Jun and p53 and repress p53 activity." *Proc Natl Acad Sci U S A* **99**(5): 2872–7.

Schwartz, D. C., R. Felberbaum, et al. (2007). "The Ulp2 SUMO protease is required for cell division following termination of the DNA damage checkpoint." *Mol Cell Biol* **27**(19): 6948–61.

Seeler, J. S., O. Bischof, et al. (2007). "SUMO, the three Rs and cancer." *Curr Top Microbiol Immunol* **313**: 49–71.

Semenza, G. L. (2007). "Evaluation of HIF-1 inhibitors as anticancer agents." *Drug Discov Today* **12**(19-20): 853–9.

Shiio, Y. and R. N. Eisenman (2003). "Histone sumoylation is associated with transcriptional repression." *Proc Natl Acad Sci U S A* **100**(23): 13225–30.

Shin, J. A., E. S. Choi, et al. (2005). "SUMO modification is involved in the maintenance of heterochromatin stability in fission yeast." *Mol Cell* **19**(6): 817–28.

Shinbo, Y., T. Taira, et al. (2005). "DJ-1 restores p53 transcription activity inhibited by Topors/p53BP3." *Int J Oncol* **26**(3): 641–8.

Shinbo, Y., T. Niki, et al. (2006). "Proper SUMO-1 conjugation is essential to DJ-1 to exert its full activities." *Cell Death Differ* **13**(1): 96–108.

Stead, K., C. Aguilar, et al. (2003). "Pds5p regulates the maintenance of sister chromatid cohesion and is sumoylated to promote the dissolution of cohesion." *J Cell Biol* **163**(4): 729–41.

Steffan, J. S., N. Agrawal, et al. (2004). "SUMO modification of Huntingtin and Huntington's disease pathology." *Science* **304**(5667): 100–4.

Strunnikov, A. V., L. Aravind, et al. (2001). "*Saccharomyces cerevisiae* SMT4 encodes an evolutionarily conserved protease with a role in chromosome condensation regulation." *Genetics* **158**(1): 95–107.

Sun, H., J. D. Leverson, et al. (2007). "Conserved function of RNF4 family proteins in eukaryotes: targeting a ubiquitin ligase to SUMOylated proteins." *Embo J* **26**(18): 4102–12.

Tanaka, K., J. Nishide, et al. (1999). "Characterization of a fission yeast SUMO-1 homologue, pmt3p, required for multiple nuclear events, including the control of telomere length and chromosome segregation." *Mol Cell Biol* **19**(12): 8660–72.

Tatham, M. H., E. Jaffray, et al. (2001). "Polymeric chains of SUMO-2 and SUMO-3 are conjugated to protein substrates by SAE1/SAE2 and Ubc9." *J Biol Chem* **276**(38): 35368–74.

Tatham, M. H., M. C. Geoffroy, et al. (2008). "RNF4 is a poly-SUMO-specific E3 ubiquitin ligase required for arsenic-induced PML degradation." Nat Cell Biol.

Taylor, D. L., J. C. Ho, et al. (2002). "Cell-cycle-dependent localisation of Ulp1, a Schizosaccharomyces pombe Pmt3 (SUMO)-specific protease." *J Cell Sci* **115**(Pt 6): 1113–22.

Ueda, H., J. Goto, et al. (2002). "Enhanced SUMOylation in polyglutamine diseases." *Biochem Biophys Res Commun* **293**(1): 307–13.

Um, J. W. and K. C. Chung (2006). "Functional modulation of parkin through physical interaction with SUMO-1." *J Neurosci Res* **84**(7): 1543–54.

Um, J. W., D. S. Min, et al. (2006). "Parkin ubiquitinates and promotes the degradation of RanBP2." *J Biol Chem* **281**(6): 3595–603.

Upadhya, S. C. and A. N. Hegde (2007). "Role of the ubiquitin proteasome system in Alzheimer's disease." *BMC Biochem* **8 Suppl 1**: S12.

Urnov, F. D., A. P. Wolffe, et al. (2001). "Molecular mechanisms of corepressor function." *Curr Top Microbiol Immunol* **254**: 1–33.

Uzunova, K., K. Gottsche, et al. (2007). "Ubiquitin-dependent proteolytic control of SUMO conjugates." *J Biol Chem* **282**(47): 34167–75.

Veltman, I. M., L. A. Vreede, et al. (2005). "Fusion of the SUMO/Sentrin-specific protease 1 gene SENP1 and the embryonic polarity-related mesoderm development gene MESDC2 in a patient with an infantile teratoma and a constitutional t(12;15)(q13;q25)." *Hum Mol Genet* **14**(14): 1955–63.

Vertegaal, A. C., S. C. Ogg, et al. (2004). "A proteomic study of SUMO-2 target proteins." *J Biol Chem* **279**(32): 33791–8.

Vogelauer, M., J. Wu, et al. (2000). "Global histone acetylation and deacetylation in yeast." *Nature* **408**(6811): 495–8.

von Mikecz, A., S. Zhang, et al. (2000). "CREB-binding protein (CBP)/p300 and RNA polymerase II colocalize in transcriptionally active domains in the nucleus." *J Cell Biol* **150**(1): 265–73.

Wang, L. and S. Banerjee (2004). "Differential PIAS3 expression in human malignancy." *Oncol Rep* **11**(6): 1319–24.

Weger, S., E. Hammer, et al. (2003). "The DNA topoisomerase I binding protein topors as a novel cellular target for SUMO-1 modification: characterization of domains necessary for subcellular localization and sumolation." *Exp Cell Res* **290**(1): 13–27.

Weger, S., E. Hammer, et al. (2005). "Topors acts as a SUMO-1 E3 ligase for p53 in vitro and in vivo." *FEBS Lett* **579**(22): 5007–12.

Wong, K. A., R. Kim, et al. (2004). "Protein inhibitor of activated STAT Y (PIASy) and a splice variant lacking exon 6 enhance sumoylation but are not essential for embryogenesis and adult life." *Mol Cell Biol* **24**(12): 5577–86.

Xie, Y., O. Kerscher, et al. (2007). "The yeast Hex3.Slx8 heterodimer is a ubiquitin ligase stimulated by substrate sumoylation." *J Biol Chem* **282**(47): 34176–84.

Yang, S. H., E. Jaffray, et al. (2003). "Dynamic interplay of the SUMO and ERK pathways in regulating Elk-1 transcriptional activity." *Mol Cell* **12**(1): 63–74.

Yang, S. H. and A. D. Sharrocks (2004). "SUMO promotes HDAC-mediated transcriptional repression." *Mol Cell* **13**(4): 611–7.

Zhong, S., S. Muller, et al. (2000a). "Role of SUMO-1-modified PML in nuclear body formation." *Blood* **95**(9): 2748–52.

Zhong, S., P. Salomoni, et al. (2000b). "The transcriptional role of PML and the nuclear body." *Nat Cell Biol* **2**(5): E85–90.P. McChesney and G.M. Kupfer (✉)

Department of Pediatrics, Section of Pediatric Hematology-Oncology,
Yale University School of Medicine, New Haven, CT, USA

Proteasome Inhibitors: Recent Progress and Future Directions

Marie Wehenkel, Yik Khuan Ho, and Kyung-Bo Kim

Abstract The proteasome continues to be an attractive target for cancer drug discovery and was validated as such by the successful development of bortezomib. Despite its remarkable efficacy, concerns regarding side effects and drug resistance limit the widespread application of bortezomib. Thus, there has been a heightened interest in the development of a new class of proteasome inhibitors. In this chapter, we discuss recent efforts towards the development of proteasome inhibitors. In addition, we discuss a novel class of compounds targeting an alternative proteasome, the immunoproteasome.

Keywords Ubiquitin-proteasome system • Proteasome inhibitors • Pharmacophores • Catalytic β-subunits • N-terminal nucleophile protease • Subunit-specific inhibitors • Immunoproteasome-specific inhibitors, Activity-based probes

Introduction

The proteasome plays a central role in maintaining intracellular protein homeostasis and regulating important signaling pathways as a final executioner in the ubiquitin-proteasome system (UPS) (Hershko and Ciechanover 1998; Myung et al. 2001a, b). Thus the use of proteasome inhibitors as molecular probes has tremendously impacted our understanding of biological processes regulated by the proteasome in cells (Kisselev and Goldberg 2001). Importantly, proteasome inhibitors have also provided a clinically valuable therapeutic agent, as highlighted by the FDA approval of bortezomib for relapsed multiple myeloma (Adams 2002). Due to the successful applications of proteasome inhibitors as therapeutic agents, major efforts have been focused on the development of proteasome inhibitors with higher therapeutic efficacy.

M. Wehenkel, Y.K. Ho, and K-B. Kim (✉)
Department of Pharmaceutical Sciences, College of Pharmacy,
University of Kentucky, Lexington, KY, USA

K. Sakamoto and E. Rubin (eds.), *Modulation of Protein Stability in Cancer Therapy*,   99
DOI: 10.1007/978-0-387-69147-3_7, © Springer Science+Business Media, LLC 2009

Particular attention has also been paid to the synthesis of highly sophisticated and specialized inhibitors designed for specific purposes, such as subunit-specific inhibitors (Kim and Crews 2003; Myung et al. 2001a, b). In recent years, there have also been increasing efforts to develop compounds that selectively inhibit catalytic β subunits of the immunoproteasome, an alternative form of the constitutive proteasome produced in mammalian cells upon exposure to stimuli, such as interferon-γ. This synthetic efforts is due in large part to recent observations that catalytic β subunits of the immunoproteasome are upregulated in certain pathological disorders but not in normal states (Ho et al. 2007; Orlowski and Orlowski 2006; Orlowski 2005). To create novel classes of inhibitors, researchers have employed both a medicinal chemistry approach and natural product screening strategy, which provide proteasome inhibitors with unusual structures or unique pharmacophores (Kim and Crews 2003).

In this chapter, we discuss the chemistry of some important classes of proteasome inhibitors, with an emphasis on the most recently developed proteasome inhibitors and a novel class of inhibitors targeting the immunoproteasome. In addition, we discuss proteasome activity-based probes derived from proteasome inhibitors that can be used to monitor proteasome activity in real-time.

Protein Degradation by the Ubiquitin-Proteasome System (UPS)

The UPS is largely responsible for the targeted degradation of intracellular proteins in eukaryotes and proteasome is a central player in the UPS pathway (Ciechanover et al. 2000). The proteasome itself is a master proteolysis machine possessing multiple catalytic activities and functioning cooperatively in the UPS to facilitate vital protein destruction process. Through controlled ubiquitination and subsequent degradation, UPS regulates many important biological processes, such as cell cycle progression, differentiation, and inflammatory responses (Verma and Deshaies 2000; King et al. 1996; Karin and Ben-Neriah 2000). Unlike other nondestructive and reversible regulatory means, acetylation or phosphorylation for instance, the UPS irreversibly alters signaling pathways through the degradation of regulatory proteins. Therefore it is the most ruthless means of regulating signaling molecules and typically serves to regulate biological processes which need to be tightly controlled. To prevent promiscuous protein degradation by the proteasome, proteolytic activity is confined to an inner compartment that is only accessible to unfolded proteins. Additionally cells have developed a protein-targeting system, tightly controlled by a set of ubiquitination enzymes (Ciechanover et al. 2000). Once proteins are multiubiquitinated by these enzymes, they are rapidly recognized and degraded by the proteasome. Ironically, these tightly controlled multistep degradation procedures also provide an opportunity to intervene in the targeted degradation system at multiple stages. Among the potential intervention sites, such as ubiquitination (E1-E3), deubiquitination (de-ubiquitinase), protein unfolding (ATPase) or proteolysis (catalytic β subunits), the 20S proteasome catalytic core will be our focus. Almost all the inhibitors of proteasome function currently available target active sites of proteasome's catalytic β subunits, whereas only a few proteasome inhibitors were designed or identified to target non-β subunits of the proteasome.

Inhibitors of the Proteasome's Catalytic β Subunits

The proteasome belongs to the N-terminal nucleophile (Ntn) family of hydrolases, which use a nucleophile located at the N-terminus of a subunit to catalyze hydrolysis reactions (Brannigan et al. 1995; Oinonen and Rouvinen 2000). It has been demonstrated in the proteasome that this nucleophile is the hydroxyl side chain of the N-terminus threonine, which is directly involved in hydrolysis of peptide bonds (Lowe et al. 1995; Groll et al. 1997, 2000) (Fig. 1). In addition, the N-terminus free amine residue of a proteasome's catalytic β subunit has been shown to play a crucial role in the catalytic activity of the proteasome by functioning as a general base (Oinonen and Rouvinen 2000). Given that the nucleophilic attack on the scissile peptide bond of the substrate by the N-terminus threonine is a critical step in the proteosomal hydrolysis reactions (Fig. 1), proteasome inhibitors have been developed primarily to target this active site. The apparent preference for targeting the catalytic β subunits can be linked to the availability of inhibitors of other proteases that have similar catalytic properties (Kim and Crews 2003). In addition, the catalytic β subunits have a well-characterized active site (Borissenko and Groll 2007), which makes the design of inhibitors more straightforward.

Previously discovered molecules which inhibit proteases having similar catalytic activity as the proteasome has been instrumental in the development of the first generation of highly potent synthetic proteasome inhibitors (Kim and Crews 2003). For example, leupeptin, a standard serine and cysteine protease inhibitor, and calpain inhibitors I and II are aldehyde-based inhibitors of proteases having catalytic activities similar to those of the proteasome. They were additionally shown to

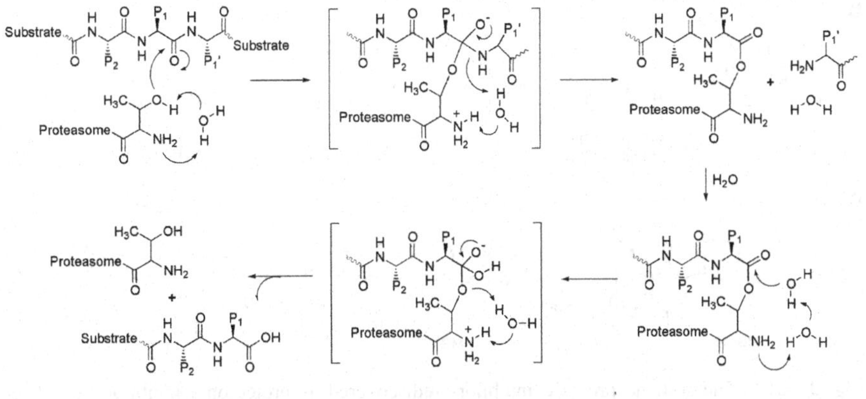

Fig. 1 The proposed proteolytic mechanism catalyzed by proteasome

possess proteasomal inhibitory activity (Wilk and Orlowski 1983a,b) (see Fig. 2a). The peptide aldehydes have been shown to inhibit proteasomal proteolysis via the formation of a reversible hemiacetal linkage between the aldehyde pharmacophore and the hydroxyl side chain of the N-terminus threonine of the catalytic β subunits (Lowe et al. 1995; Groll et al. 1997) (Fig. 2b). The rediscovery of these protease inhibitors as proteasome inhibitors accelerated efforts towards the development of more potent and specific peptide aldehyde inhibitors, such as MG115 and MG132 (Rock et al. 1994) (Fig. 2c). Similar to these aldehyde-based inhibitors, almost all the synthetic proteasome inhibitors developed to date are composed of two distinct moieties. They are usually comprised of a peptide or peptide-like backbone, which provides the necessary binding affinity and specificity towards the proteasome. Additionally they contain an eletrophilic "warhead" having the ability to undergo nucleophilic attack, which is spearheaded by the hydroxyl side chain of N-terminus threonine of the catalytic β subunit. Although there have been increasing efforts to target novel moieties for proteasome inhibition, the catalytic β subunits of the proteasome remain the most attractive and effective target to block protein degradation.

Several additional electrophilic groups have been employed as a warhead in the design of proteasome inhibitors, exploiting the nucleophilic attack mechanism by the hydroxyl side chain of the N-terminus threonine of the catalytic β subunit. Depending on the electrophilic warhead utilized, a reversible or irreversible complex is formed with the hydroxyl group of the N-terminus threonine of the catalytic β subunits. While synthetic approaches produced exclusively peptide-based proteasome inhibitors, natural products have provided proteasome inhibitors with unusual pharmacophores or nonpeptidic molecular skeletons (for review, see Kim and Crews 2003). Almost

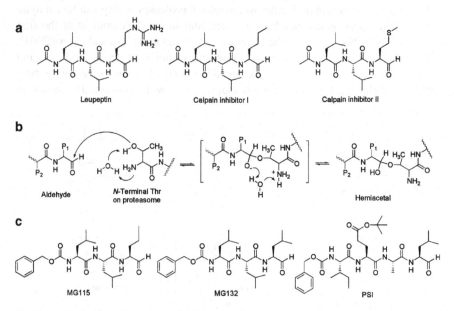

Fig. 2 (**a**) Serine/cysteine protease inhibitors rediscovered as proteasome inhibitors; (**b**) The proposed inhibitory mechanism of aldehyde pharmacophore; (**c**) Synthetic peptide aldehydes designed to target proteasome

all of these synthetic and natural product inhibitors block the N-terminus active sites of multiple catalytic β subunits in the proteasome and thus are classified as "broad spectrum" proteasome inhibitors.

"Broad Spectrum" Proteasome Inhibitors

The proteasome contains three major proteolytic activities which are generally classified as chymotrypsin-like (CT-L), trypsin-like (T-L), and caspase-like (C-L) (Cardozo 1993; Orlowski 1993). These names refer to the proteolytic mechanisms utilized by each catalytic β subunit, so CT-L cleaves after aromatic or bulky hydrophobic amino acids, T-L cleaves after basic residues, and C-L cleaves after acidic residues. Extensive kinetic studies, combined with X-ray data analysis, revealed that different proteasome β subunits could be assigned different catalytic activities (Lowe et al. 1995; Groll et al. 1997). For example, X/β5 is responsible for the CT-L activity, while Y/β1 and Z/β2 are responsible for C-L and T-L activities, respectively (for review, see Myung et al. 2001a, b). Among these catalytic activities, it has been shown that the CT-L activity is largely responsible for the proteolytic activity of the proteasome and thus, its biological functions in vivo and in vitro (Myung et al. 2001a, b; Kisselev et al. 2006; Figueiredo-Pereira et al. 1994, 1996).

Almost all the proteasome inhibitors developed or identified to date are not particularly specific towards one type of catalytic β subunit of the proteasome, but rather target at least two different types of catalytic β subunits. The first broad-spectrum proteasome inhibitors that were widely used as molecular probes are peptide aldehydes, specifically, MG132 and MG115 (see Fig. 2b), tripeptide aldehydes developed by Rock and colleagues (Rock et al. 1994). Over the years, these peptide aldehydes assisted in the identification of signaling molecules and pathways regulated by the UPS. Although these peptide aldehydes are sufficiently potent CT-L activity inhibitors, they are not suitable for dissecting the contributions of the proteasome's CT-L activity since they also inhibit the other major catalytic activities of the proteasome. To improve the potency and specificity of the aldehyde pharmacophore-based inhibitors towards the CT-L activity of the proteasome, additional efforts have been made to develop more specific and potent inhibitors. One such example is the development of PSI (Wilk and Figueiredo-Pereira 1993), a linear peptide aldehyde (see Fig. 2b). PSI is a highly potent CT-L activity inhibitor prepared by Wilk and colleagues and showed good membrane permeability. Overall, despite the peptide aldehydes having long been employed as useful molecular probes, any additional applications are still limited due to concerns about off-target activity. For the same reason, the electrophilic aldehyde warhead also has a limited potential as a pharmaceutical agent.

Since the development of peptide aldehydes, proteasome inhibitors have been aggressively pursued via a synthetic approach (for review, see Kim and Crews 2003). For example, Bogyo et al. (1997, 1998) successfully developed a class of peptide vinyl sulfone-based proteasome inhibitors (Fig. 3a). The vinylsulfone pharmacophore was originally introduced as a mechanism-based cysteine protease inhibitor by Palmer and colleagues but it also acts as a Michael acceptor for the nucleophilic hydroxyl side chain of the N-terminus threonine of catalytic β subunits (Bromme et al. 1996; Palmer et al. 1995). By acting as a Michael acceptor, the vinyl sulfone pharmacophore forms a covalent bond (an ether linkage) with the hydroxyl group in the active site of the proteasome (Fig. 3b). Later, Kessler et al developed a long hydrocarbon-containing peptide vinyl sulfone that inhibits all three active sites equally well, thereby providing an additional tool for UPS studies (Kessler et al. 2001; van Swieten et al. 2007). Nevertheless, the lack of specificity is a major concern for this class of inhibitors, since the peptide vinyl sulfones inhibit both the proteasome and cysteine proteases.

The boronic acid-based peptides are another family of synthetic broad-spectrum proteasome inhibitors that target the N-terminus active site of catalytic β subunits (Fig. 4). Initially, the peptide boronic esters were developed to target serine proteases,

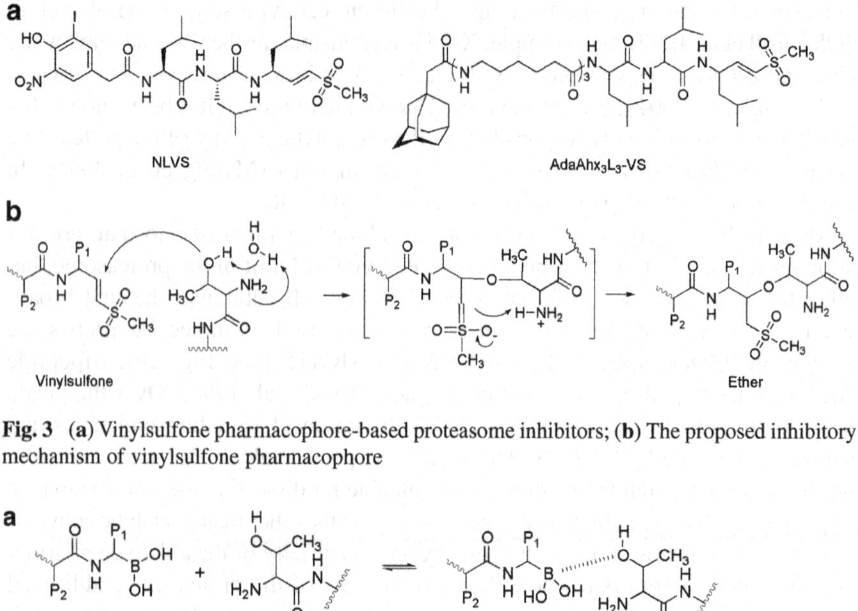

Fig. 3 (**a**) Vinylsulfone pharmacophore-based proteasome inhibitors; (**b**) The proposed inhibitory mechanism of vinylsulfone pharmacophore

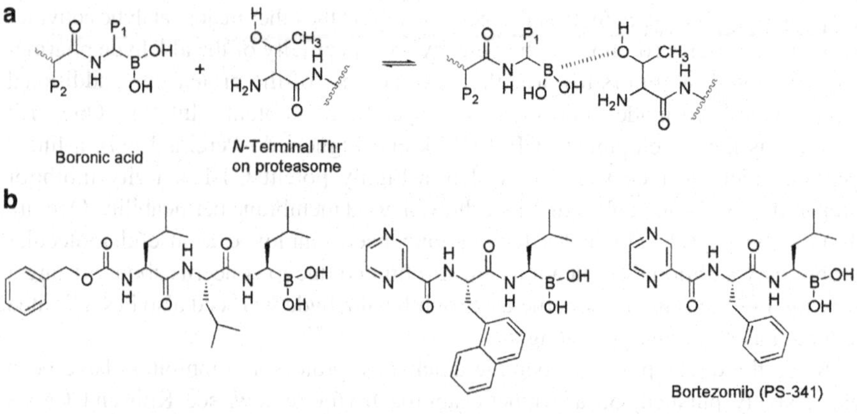

Fig. 4 (**a**) The proposed inhibitory mechanism of the boronic acid pharmacophore; (**b**) Boronic acid pharmacophore-based proteasome inhibitors

such as thrombin (Fevig et al. 1998). A strong interaction between these proteases and the boronic ester is thought to be due to a complex formation between the boron and the hydroxyl group of the N-terminus. This occurs when an empty *p*-orbital on a boron atom interacts with the oxygen lone pair of the serine hydroxyl side chain in the active site, giving a stable pseudo-tetrahedral complex (Fig. 4a). This unique interaction between boron and oxygen is thought to provide the selectivity of boronic esters for serine proteases over cysteine proteases (Fevig et al. 1998), as the thiol group of cysteine proteases provides a relatively weak bonding interaction with boron. Using this boronic acid-based pharmacophore, Adams et al. have prepared potent di- and tripeptidyl inhibitors (Adams et al. 1998) (Fig. 4b). Eventually, they developed a potent dipeptidyl boronic acid proteasome inhibitor, bortezomib (PS-341), which was approved by the FDA as Velcade™ for the treatment of relapsed multiple myeloma (Adams 2002). In recent years, bortezomib has been widely used not only as a successful chemotherapeutic agent but also as a molecular probe of disease states. Unfortunately, the broad application of bortezomib is currently limited due to drug-associated side effects. In this regard, it has been reported that bortezomib is oxidatively metabolized via the cytochrome P450 enzymes and the boronic acid residue of bortezomib is released to yield inactive boron-less bortezomib (Lu et al. 2006). This is perhaps one of the mechanisms of bortezomib clearance in the body. Currently, due to the development of resistance to the boronic acid-based bortezomib by certain types of multiple myeloma, proteasome inhibitors having other pharmacophores are being investigated for their activity in clinical trials (Joazeiro et al. 2006).

In addition to synthetic approaches, natural products have been a source of several broad-spectrum proteasome inhibitors (Myung et al. 2001a, b; Kim and Crews 2003). Traditionally, nature has been an important source of biologically active compounds (Tan et al. 2006; Harvey 1999) and thus, it is not surprising that natural products have also provided a number of proteasome inhibitors. Unlike synthetic molecules, natural products have often provided proteasome inhibitors with nonpeptide backbones and novel pharmacophores. One such natural product is lactacystin (Fig. 5a), a proteasome inhibitor with a β-lactone pharmacophore isolated as a metabolite of *Streptomyces lactacystinaeus* (Omura et al. 1991). Lactacystin was discovered because of its ability to induce neutrite outgrowth in the murine neuroblastoma cell line Neuro-2a. Fenteany et al showed that lactacystin targets the 20S proteasome by a covalent modification of the amino terminal threonine of β-subunits (Fenteany et al. 1995) (Fig. 5b). Later it was demonstrated that the active component of lactacystin is the *clasto*-lactacystin β-lactone that is a rearrangement product of lactacystin in aqueous conditions (Dick et al. 1996, 1997). Lactacystin has been shown to target multiple catalytic β subunits of the proteasome (Fenteany et al. 1995). Furthermore, although lactacystin was initially thought to be highly specific towards the proteasome, later studies revealed that lactacystin also inhibits other cellular proteases (Ostrowska et al. 1997, 2000). Despite this off-target issue, lactacystin is still one of the most widely used proteasome inhibitors in laboratories worldwide. Salinosporamide A (NPI-0052) and salinosporamide B are β-lactone-pharmacophore containing proteasome inhibitors derived from a natural source

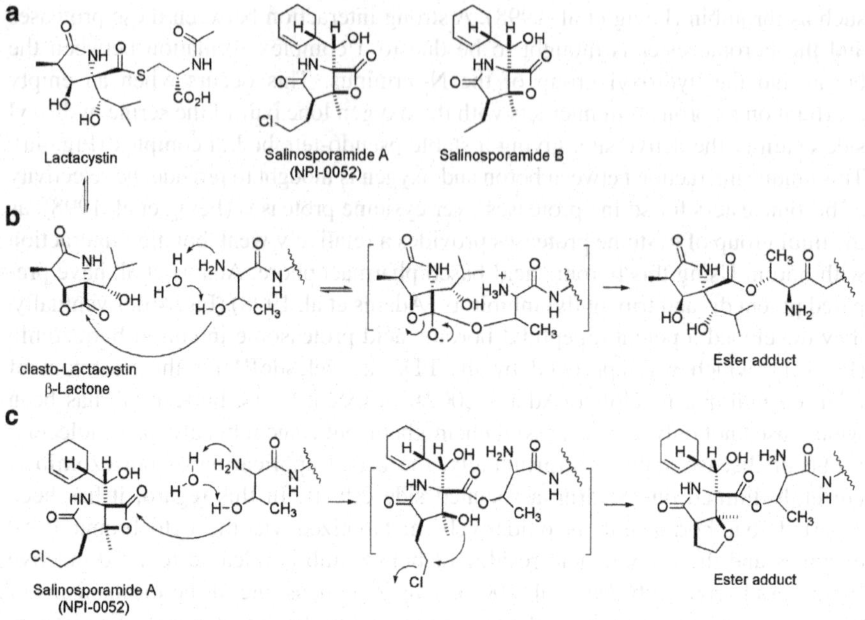

a

Lactacystin   Salinosporamide A   Salinosporamide B
(NPI-0052)

b

clasto-Lactacystin
β-Lactone   Ester adduct

c

Salinosporamide A
(NPI-0052)   Ester adduct

Fig. 5 (**a**) β-Lactone-pharmacophore-based natural product proteasome inhibitors; (**b, c**) The proposed proteasomal inhibitory mechanism of β-lactone pharmacophore

(Feling et al. 2003) (Fig. 5a). They were isolated from a species of marine *Actinomycete* and in particular, NPI-0052 is currently being developed as a promising therapeutic agent (Chauhan et al. 2006; Ruiz et al. 2006). X-ray structural studies revealed that NPI-0052 is covalently bound to the N-terminus threonine hydroxyl group of catalytic β-subunits via an ester linkage to the carbonyl derived from the β-lactone ring of NPI-0052 (Groll et al. 2006) (Fig. 5c). Interestingly, NPI-0052 has been shown to be effective against strains of multiple myeloma that have developed resistance to bortezomib (Chauhan et al. 2004).

Another important class of proteasome inhibitors arising from natural products is a family of linear peptide epoxyketones (Fig. 6a). Although the α′,β′-epoxyketone pharmacophore was initially designed in the preparation of synthetic proteasome inhibitors (Spaltenstein et al. 1996), their selectivity and pharmacological importance was not realized until the successful mode of action studies of two such natural products, epoxomicin and eponemycin, by Crews et al (Sin et al. 1998, 1999; Meng et al. 1999a, b). The antitumor natural product epoxomicin is a linear peptide α′,β′-epoxyketone isolated from an unidentified species of *Actinomycete* (strain No. Q996-17) (Hanada et al. 1992). The elegant mode of action studies performed by Crews and colleagues showed that epoxomicin targets the 20S proteasome with unusually high specificity (Groll et al. 2000; Sin et al. 1999; Meng et al. 1999a, b). X-ray structural studies suggested that the high specificity towards the proteasome is due to the formation of an unusual 6-membered morpholino ring between the amino terminal catalytic threonine of the 20S proteasome and the α′,β′-epoxyketone

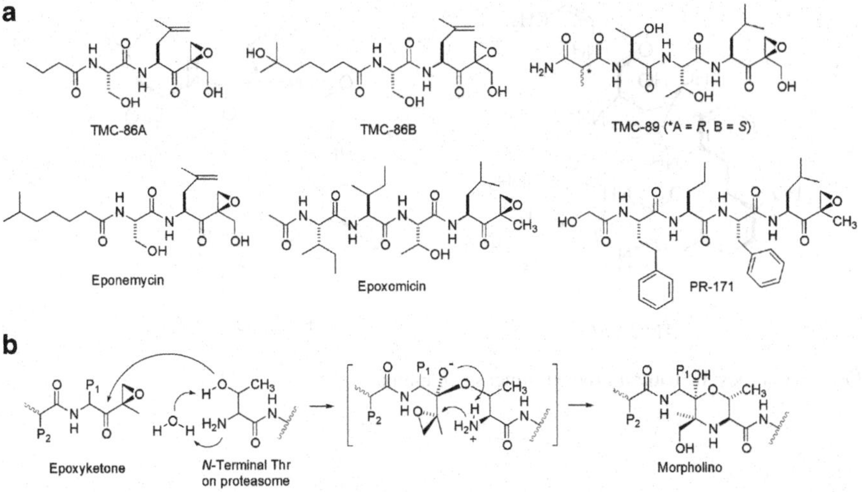

a

TMC-86A   TMC-86B   TMC-89 (*A = R, B = S)

Eponemycin   Epoxomicin   PR-171

b

Epoxyketone   N-Terminal Thr on proteasome   Morpholino

Fig. 6 (**a**) Epoxyketone pharmacophore-based proteasome inhibitors isolated or derived from natural products; (**b**) The proposed inhibitory mechanism of epoxyketone pharmacophore

pharmacophore of epoxomicin (Groll et al. 2000) (Fig. 6b). An antiangiogenic natural product, eponemycin, is another example of a linear peptide α',β'-epoxyketone isolated from *Streptomyces Hygroscopicus* (No P247-271) on the basis of its activity against B16 melanoma (Sugawara et al. 1990). Crews et al showed that, like epoxomicin, eponemycin also target the 20S proteasome with high specificity (Meng et al. 1999a, b). In addition to epoxomicin and eponemycin, natural products have provided a number of additional linear peptide α',β'-epoxyketone proteasome inhibitors (Koguchi et al. 2000). Furthermore, synthetic approaches based on the epoxyketone peptide skeleton of epoxomicin have yielded several useful molecules. One of them, PR-171, is currently being investigated in clinical trials for its efficacy against multiple myeloma (Demo et al. 2007).

The discovery and use of these natural products as proteasome inhibitors has invigorated a systematic natural product screening approach for novel proteasome inhibitors. One advantage of a natural product screening approach is that it often provides biologically active compounds with a nonpeptidic backbone. Not unexpectedly, a number of proteasome inhibitors with unusual structures have been identified through systematic screening. Among these inhibitors, the most interesting compounds are a series of macrocyclic molecules isolated from the fermentation broth of *Apiospora montagnei Sacc.* (TC 1093), such as TMC-95 (Koguchi et al. 2000; Kohno et al. 2000) (Fig. 7). Despite their macrocyclic features, these compounds possess potent inhibitory activity towards the CT-L activity of the proteasome. Meanwhile, they do not inhibit other proteases, such as *m*-calpain, cathepsin L, and trypsin. X-ray crystal data analysis of the yeast 20S proteasome:TMC-95 complex revealed that multiple hydrogen-bond networks between the main-chain atoms of the proteasome and TMC-95 are major contributors towards the high affinity binding of TMC-95 with

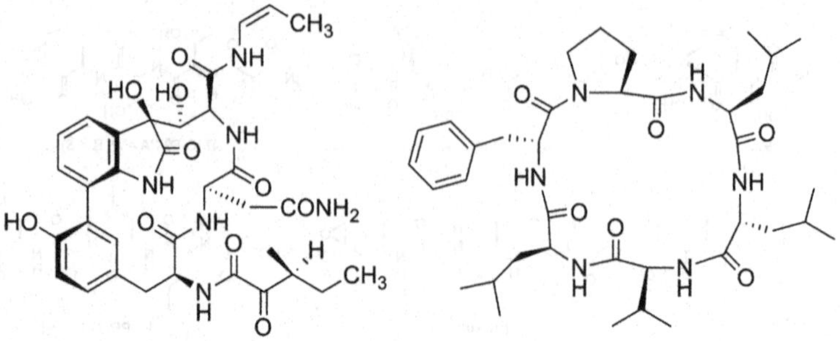

TMC-95A   Phepropeptin A

Fig. 7 Macrocyclic natural product proteasome inhibitors

the proteasome (Groll et al. 2001). In addition, phepropeptin A, a cyclic hexapeptide molecule isolated from *Streptomyces sp.* MK600-cF7 in the course of natural product screening, has also been shown to inhibit the proteasome (Sekizawa et al. 2001) (Fig. 7). As compared to linear peptide backbone-based inhibitors, one advantage of these natural products with unusual skeletons is their stability in cells due to high resistance to intracellular proteases. It will be interesting to see if these natural products can be further developed for the treatment of cancers.

Traditional medicine, which employs a variety of natural materials such as remedies for illness and disease, has been used for centuries in many regions of the world. While most of these medicines have been proven safe and effective, their mechanisms of action are often not clearly understood. Considering that inhibition of the proteasome results in a variety of biological activity commonly produced by traditional medicine, such as antitumor or antiinflammatory properties, it is expected that the active components of some traditional medicines will have proteosomal inhibitory activity. In fact, epigallocatechin-3-gallate (EGCG) (Figure 8a), one of the major component of green tea, has been recognized for its antitumor and antiinflammatory activities, while showing itself to be a potent inhibitor of the CT-L activity of the 20S proteasome (Nam et al. 2001; Chen et al. 2004). Careful analysis of the limited atomic orbital and SAR (structure-activity-relationship) studies as well as HPLC data of EGCG reaction products with the purified proteasome suggested that a nucleophilic attack by the N-terminal threonine of the proteasome occurs at the ester bond located between the two aromatic residues (Fig. 8b). In addition, a number of chemopreventive or antitumor dietary flavonoids and triterpenoids, such as genistein and celastrol (Fig. 9), have been reported to possess proteasome inhibitory activity (Chen et al. 2004, 2005; Yang et al. 2006, 2007; Kazi et al. 2003). While these dietary chemicals have been shown to primarily inhibit the CT-L activity of the proteasome in vitro, it remains to be investigated whether the proteasome mediates part or all of the pharmacological activities of these natural products in vivo. Mode of action studies on these dietary compounds will aid our understanding of the mechanism of their protective properties.

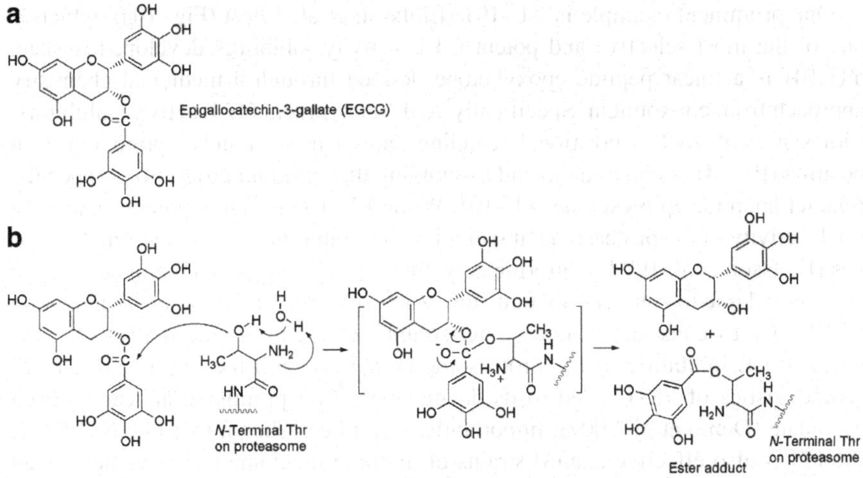

Fig. 8 (**a**) EGCG, a major component of green tea; (**b**) The proposed mechanism of proteasome inhibition by EGCG

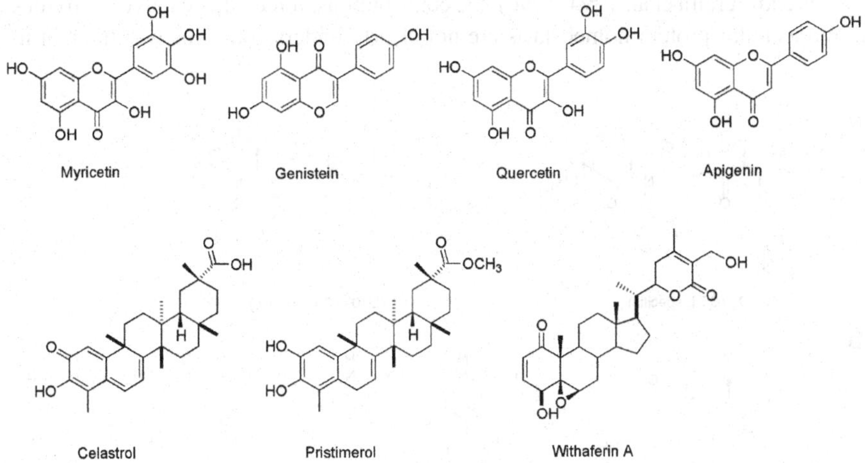

Fig. 9 Dietary phytochemicals whose mode of action has been partially attributed to proteasome inhibition

Subunit-Specific Inhibitors of the Proteasome

Although most natural and synthetic proteasome inhibitors primarily target the CT-L activity of proteasome, they still inhibit the other catalytic activities. Since the CT-L activity of the proteasome is largely responsible for protein degradation in cells, researchers have focused on the development of proteasome inhibitors that exclusively block CT-L activity or the subunit responsible for CT-L activity. Such a molecule would have enhanced potential as a pharmaceutical agent due to its increased specificity and perhaps, lower toxicity.

One prominent example is YU-101 (Elofsson et al. 1999) (Fig. 10a), which is one of the most selective and potent CT-L activity inhibitors developed to date. YU-101 is a linear peptide epoxyketone derived through a medicinal chemistry approach from epoxomicin. Specifically, to develop potent CT-L activity inhibitors, Elofsson et al. took a positional scanning approach, separately optimizing each position (P1-P4) of epoxomicin and assembling the optimum groups to produce the final tetrapeptide epoxyketone, YU-101. While YU-101 is highly potent against the CT-L activity of the proteasome, it still minimally inhibits other proteasomal activities (Elofsson et al. 1999). Unfortunately, the use of YU-101 as a therapeutic agent was limited due to concerns about its low water solubility. Eventually, optimization of YU-101 to construct a more water-soluble analogue produced PR-171. This potent CT-L inhibition of the proteasome combined with the improved delivery characteristics of PR-171 led to its development as a promising anticancer drug candidate (Demo et al. 2007). Importantly, it has been shown that like NPI-0052, PR-171 is also effective against strains of multiple myeloma that have developed resistance to bortezomib (Demo et al. 2007).

While the CT-L activity of the proteasome was shown to be largely responsible for protein degradation in vivo and in vitro (Myung et al. 2001a, b; Kisselev et al. 2006; Figueiredo-Pereira et al. 1994, 1996), the contributions of the other catalytic activities in intracellular protein homeostasis are not clearly understood. This was thought to

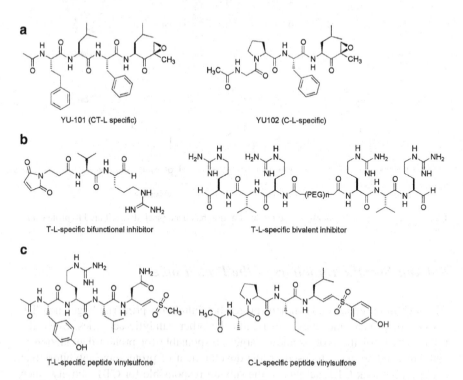

Fig. 10 Subunit-specific proteasome inhibitors with different pharmacophores; (**a**) epoxyketone; (**b**) Aldehyde; (**c**) Vinylsulfone

be due to the lack of inhibitors selective for the C-L or T-L activities. In an attempt to investigate these functions, researchers developed inhibitors that specifically target the C-L or T-L activities of the proteasome by incorporating the appropriate amino acid residues into the peptide backbone. For example, bi-functional and bi-valent T-L activity-specific inhibitors were developed based on aldehyde pharmacophore (Loidl et al. 1999a, b, 2000) (Fig. 10b). Loidl et al. incorporated a basic amino acid residue (arginine) into the P1 position of peptide aldehyde inhibitors to direct their specificity to T-L activity-responsible catalytic β subunits. While they are highly specific for the T-L activity-responsible catalytic β subunit, the protein-like long structural features and off-target issues make them undesirable for use as a molecular probe in biological studies. Meanwhile, Bogyo and Nazif developed T-L activity-specific compounds using peptide vinyl sulfone pharmacophore (Nazif and Bogyo 2001) (Fig. 10c). On the other hand, Crews and colleagues developed a series of C-L activity-specific peptide α′,β′-epoxyketones by optimizing the amino acids at the P1-P4 positions of epoxomicin (Myung et al. 2001a, b) (Fig. 10a). They used the C-L inhibitor (YU-102) to show that selective inhibition of the C-L activity was not sufficient to inhibit total protein degradation in living cells (Myung et al. 2001a, b). Recently, Kisselev and colleagues developed a cell-permeable peptide vinyl sulfone inhibitor of the proteasome's C-L activity site (van Swieten et al. 2007). Despite recent developments in subunit-specific inhibitors, the functions of the non-CT-L activities and the extent to which these activities contribute to protein degradation in vivo remains to be determined. Overall, common concerns for these subunit-specific inhibitors are off-target issues and their relatively short duration of action, which makes it difficult to apply them over physiologically relevant periods of time.

Inhibitors of Alternative Proteasomes: Immunoproteasome-Specific Inhibitors

In addition to the constitutive proteasome, mammalian cells produce an alternative proteasome form called the "immunoproteasome" in response to stimuli such as interferon-γ. The immunoproteasome is composed of inducible catalytic subunits (LMP2/β1i, MECL1/β2i and LMP7/β5i) which replace their constitutive proteasome counterparts Y/β1, Z/β2 and X/β5 (Kloetzel 2001). It is widely believed that, when compared to the constitutive proteasome, the immunoproteasome has an enhanced capacity to produce peptides more suitable for MHC class I presentation because of its reduced ability to produce peptides bearing acidic residues at their C-termini. Thus it was initially suggested that the primary function of the immunoproteasome was efficient generation of MHC Class I peptides.

Recent studies increasingly suggest that the roles of the immunoproteasome in cells are far more multifaceted than what was previously believed (Yewdell 2005). Specifically, elevated expression level of catalytic subunits of immunoproteasome has been shown to be correlated with certain pathological disorders, such as neurodegerative diseases, inflammatory bowel disease (IBD), and certain types of cancer (Fitzpatrick et al. 2006, 2007). For example, hematological cancers have been shown to express elevated levels of immunoproteasome catalytic subunits (Orlowski and Orlowski 2006; Orlowski 2005). It was also reported that certain subtypes of prostate cancer cell lines express an elevated level of immunoproteasome catalytic subunits (Ho et al. 2007). All of these observations have led to the reasonable deduction that the immunoproteasome plays a key role in survival of these cells and thus, selective inhibition of the immunoproteasome may be an attractive strategy for potential therapeutic intervention. Thus far, the extent to which the immunoproteasome contributes to protein homeostasis or regulation of signaling pathways in normal or cancerous cells is not known. This is due in large part to the lack of appropriate molecular probes, such as immunoproteasome-specific inhibitors, to study such relationships in these systems. Moreover, it is expected that immunoproteasome-specific compounds will not only provide a tool to validate immunoproteasome as a chemotherapeutic target but will also serve a dual role as mechanistic probes of immunoproteasome's function in these disease states.

For this reason, increasing efforts have been made to develop immunoproteasome-specific inhibitors to test whether the immunoproteasome is a valid target for the treatment of cancerous malignancies. Orlowski et al have developed an aldehyde-based immunoproteasome inhibitor (Cbz-LnL-CHO) primarily targeting hematological cancers through the inhibition of the CT-L activity of the immunoproteasome (Orlowski and Orlowski 2006; Orlowski 2005) (Fig. 11a). However, detailed information on the biological effects of these peptide aldehydes in cells has not been reported. Similarly, Proteolix, Inc. (South San Francisco) developed peptide epoxyketone-based proteasome inhibitors that preferentially target LMP7 (Fig. 11a), one of major catalytic β subunit of the immunoproteasome (personal communication, Mike Bennet). More recently, Ho et al. have developed an LMP2/β1i-specific inhibitor (UK-101) designed based on the highly proteasome-specific epoxyketone pharmacophore (Ho et al. 2007) (Fig. 11b, c). In addition to cells of lymphatic origin, they have observed that aggressive, metastatic prostate cancer cell lines express a higher level of LMP2, as compared to benign prostate cancer cell lines. The differential expression of LMP2 is shown to be correlated with the increased sensitivity of these cells to the LMP2 inhibitor UK-101 (Ho et al. 2007). Such results indicate the possibility of expanding the immunoproteasome targeting strategy from hematological cancers to solid tumors, such as prostate cancer. An important implication of this development is that selective immunoproteasome inhibitors provide an opportunity for therapeutic intervention with significantly lower toxicity because they don't target a process essential for the survival of all eukaryotic cells, as broad spectrum inhibitors do.

Proteasome Inhibitors as Activity-Based Probes

Despite many problems such as selectivity issues resulting in off-target effects, some proteasome inhibitors have been valuable for the identification of key signaling proteins or biological processes regulated by the proteasome. However, these inhibitors are not designed to capture the real-time activity of the proteasome in disease or normal states. This is due to the fact that protein expression levels are not always

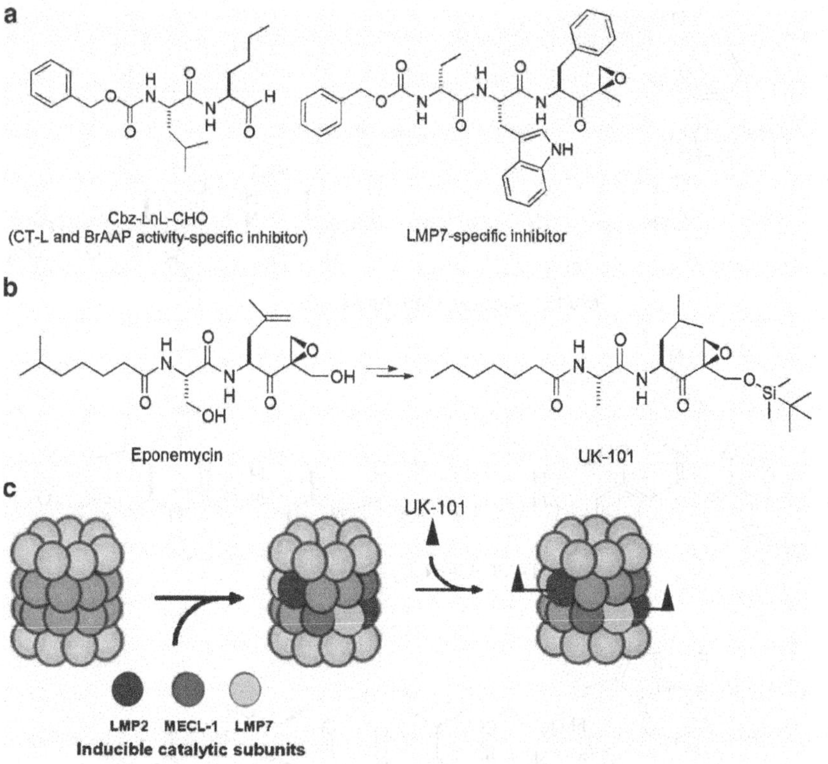

Fig. 11 (a) Eletrophilic pharmacophore-based inhibitors of immunoproteasomes; (b) Structure of UK-101, an eponemycin-derived inhibitor of immunoproteasome catalytic subunit LMP2; (c) Selective inhibition of the immunoproteasome LMP2 catalytic subunit by UK-101

directly correlated to the functional activities of proteins. The real-time monitoring of proteasomal activity in cellular models of disease states may provide important information about real-time protein homeostasis and disease progression. Accordingly, use of proteasome inhibitors has evolved applications as active site-directed activity-based probes, allowing for the measurement, in real time, of proteasomal activity. To do this, researchers have prepared broadly acting proteasome inhibitors with fluorescent tags (Verdoes et al. 2006a, b, 2007; Berkers et al. 2005, 2007).

One such compound is MV151, developed by Verdoes et al (Verdoes et al. 2006a, b). MV151 (Bodipy TMR-Ahx$_3$L$_3$VS) binds to all catalytic subunits of the proteasome in living cells and is a good example of a fluorescent activity probe (Verdoes et al. 2006a, b) (Fig. 12). Its potential uses can include the clinical profiling of proteasome activity, biochemical analysis of the subunit specificity of inhibitors, and biological analysis of proteasome functions and dynamics. The application of proteasome inhibiting fluorescent probes is reminiscent of the activity-based probes targeting cysteine proteases developed by Bogyo et al (Sadaghiani et al. 2007; Blum et al. 2005). To prepare more effective activity-based probes, researchers have synthesized fluorescent reagent-tagged proteasome inhibitors having different

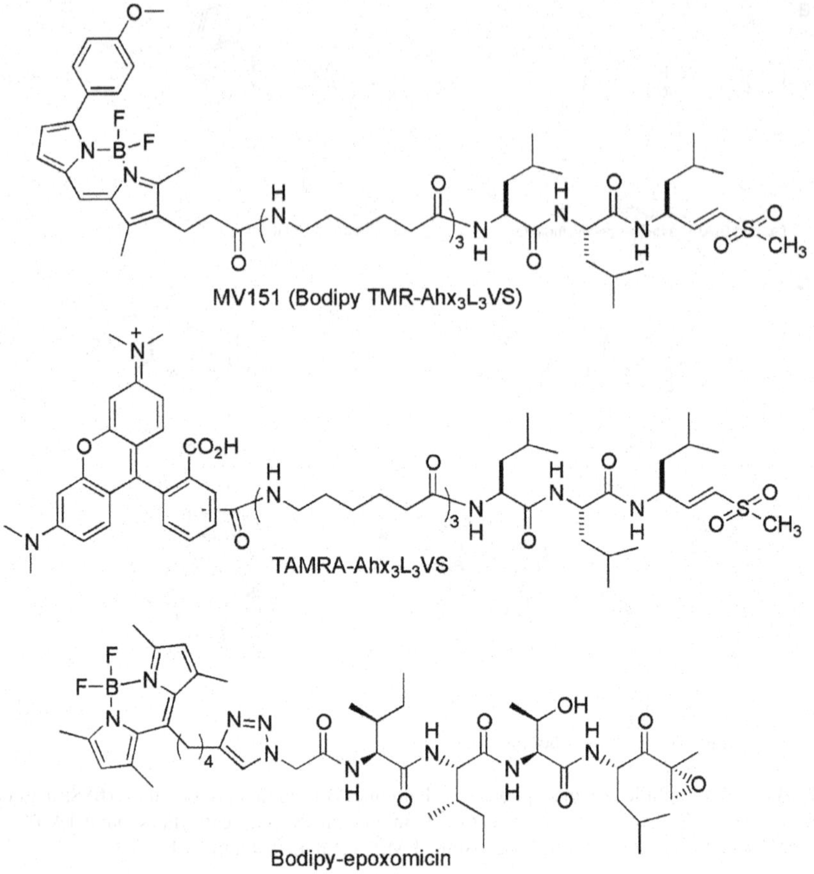

Fig. 12 Proteasome inhibitor-based fluorescent probes

irreversible pharmacophore or fluorescent molecules (Berkers et al. 2007) (Fig. 12). In addition, a natural product proteasome inhibitor epoxomicin has been used for the synthesis of an activity probe that can be used to measure levels of catalytically active proteasome in real-time (Verdoes et al. 2007) (Fig. 12). These kinds of molecules will be highly useful to show that functional proteasome plays a critical role in protein degradation in living cells and is correlated with disease progression.

Future Directions

The proteasome is a validated target for cancer treatment and a proteasome inhibitor (bortezomib) has been approved by FDA for the treatment of relapsed multiple myeloma. Despite successful therapeutic application of bortezomib, drug-associated side effects still remain a major concern, as the proteasome is essential for all cells in the body. Thus, considerable and sustained efforts have been made to develop proteasome inhibitors with lower toxicity.

In order to achieve higher specificity and thus increase the efficacy of therapeutic agents targeting the UPS, targeting upstream of proteasome degradation may be desirable, i.e., ubiquitination by E3 ligase or protein unfolding catalyzed by ATPase. However, this strategy has not been exploited. An additional untested strategy is to target catalytically nonactive sites of the 26S proteasome. Although such strategies indirectly inhibit proteasome function and are not yet tested for therapeutic intervention, interest in noncatalytic site-targeting inhibitors may continue due to the necessity for cancer drugs that may have better efficacy.

An additional approach to overcome the drug-associated side effects of broad spectrum proteasome inhibitors is to selectively inhibit the proteasome in disease states but not in normal cells of the body. For this reason, the immunoproteasome has drawn considerable attention, since abnormally elevated levels of immunoproteasome catalytic subunits are seen in certain types of cancers. Although primary function of immunoproteasome is thought to be generation of antigenic peptides, recent studies raise the possibility that MHC class I antigen processing may not be the primary function of immunoproteasome. Therefore, the immunoproteasome is not only a promising therapeutic target but also an interesting mechanistic target for the study of proteasomal biology. Currently, a major obstacle for the burgeoning field of immunoproteasome research is the lack of immunoproteasome-specific inhibitors that can be used as molecular probes and help to validate immunoproteasome as a drug discovery target. Given the successful development of immunoproteasome-specific inhibitors by several groups, the next couple of years will be exciting as these compounds undoubtedly will help to further elucidate the biological functions of the immunoproteasome. Furthermore, the potential of UPS as an antioncogenic target is just beginning to be appreciated, as the proteasome has been a common target for many dietary chemicals now known to be chemopreventive agents.

In addition to providing therapeutic agents, broad spectrum proteasome inhibitors have served as important molecular probes over the past decade. However, more sophisticated inhibitors such as subunit-specific inhibitors may be necessary to dissect complex signaling pathways regulated by proteasome in disease states. Continued efforts towards the development of novel proteasome inhibitors will provide us with better tools to dissect the role of proteasomes in disease states and more effective small molecules to treat diseases.

References

Adams, J. (2002). Development of the proteasome inhibitor PS-341. Oncologist 7, 9–16.

Adams, J., Behnke, M., Chen, S., Cruickshank, A.A., Dick, L.R., Grenier, L., Klunder, J.M., Ma, Y.T., Plamondon, L., and Stein, R.L. (1998). Potent and selective inhibitors of the proteasome: dipeptidyl boronic acids. Bioorg Med Chem Lett 8, 333–338.

Berkers, C.R., Verdoes, M., Lichtman, E., Fiebiger, E., Kessler, B.M., Anderson, K.C., Ploegh, H.L., Ovaa, H., and Galardy, P.J. (2005). Activity probe for in vivo profiling of the specificity of proteasome inhibitor bortezomib. Nat Methods 2, 357–362.

Berkers, C.R., van Leeuwen, F.W., Groothuis, T.A., Peperzak, V., van Tilburg, E.W., Borst, J., Neefjes, J.J., and Ovaa, H. (2007). Profiling proteasome activity in tissue with fluorescent probes. Mol Pharm 4, 739–748.

Blum, G., Mullins, S.R., Keren, K., Fonovic, M., Jedeszko, C., Rice, M.J., Sloane, B.F., and Bogyo, M. (2005). Dynamic imaging of protease activity with fluorescently quenched activity-based probes. Nat Chem Biol 1, 203–209.

Bogyo, M., McMaster, J.S., Gaczynska, M., Tortorella, D., Goldberg, A.L., and Ploegh, H. (1997). Covalent modification of the active site threonine of proteasomal beta subunits and the Escherichia coli homolog HslV by a new class of inhibitors. Proc Natl Acad Sci U S A 94, 6629–6634.

Bogyo, M., Shin, S., McMaster, J.S., and Ploegh, H.L. (1998). Substrate binding and sequence preference of the proteasome revealed by active-site-directed affinity probes. Chem Biol 5, 307–320.

Borissenko, L., and Groll, M. (2007). 20S proteasome and its inhibitors: crystallographic knowledge for drug development. Chem Rev 107, 687–717.

Brannigan, J.A., Dodson, G., Duggleby, H.J., Moody, P.C., Smith, J.L., Tomchick, D.R., and Murzin, A.G. (1995). A protein catalytic framework with an N-terminal nucleophile is capable of self-activation. Nature 378, 416–419.

Bromme, D., Klaus, J.L., Okamoto, K., Rasnick, D., and Palmer, J.T. (1996). Peptidyl vinyl sulfones: a new class of potent and selective cysteine protease inhibitors. Biochem J 315, 85–89.

Cardozo, C. (1993). Catalytic components of the bovine pituitary multicatalytic proteinase complex (proteasome). Enzyme Protein 47, 296–305.

Chauhan, D., Li, G., Podar, K., Hideshima, T., Mitsiades, C., Schlossman, R., Munshi, N., Richardson, P., Cotter, F.E., and Anderson, K.C. (2004). Targeting mitochondria to overcome conventional and bortezomib/proteasome inhibitor PS-341 resistance in multiple myeloma (MM) cells. Blood 104, 2458–2466.

Chauhan, D., Hideshima, T., and Anderson, K.C. (2006). A novel proteasome inhibitor NPI-0052 as an anticancer therapy. Br J Cancer 95, 961–965.

Chen, D., Daniel, K.G., Kuhn, D.J., Kazi, A., Bhuiyan, M., Li, L., Wang, Z., Wan, S.B., Lam, W.H., Chan, T.H., and Dou, Q.P. (2004). Green tea and tea polyphenols in cancer prevention. Front Biosci 9, 2618–2631.

Chen, D., Daniel, K.G., Chen, M.S., Kuhn, D.J., Landis-Piwowar, K.R., and Dou, Q.P. (2005). Dietary flavonoids as proteasome inhibitors and apoptosis inducers in human leukemia cells. Biochem Pharmacol 69, 1421–1432.

Ciechanover, A., Orian, A., and Schwartz, A.L. (2000). Ubiquitin-mediated proteolysis: biological regulation via destruction. Bioessays 22, 442–451.

Demo, S.D., Kirk, C.J., Aujay, M.A., Buchholz, T.J., Dajee, M., Ho, M.N., Jiang, J., Laidig, G.J., Lewis, E.R., Parlati, F., Shenk, K.D., Smyth, M.S., Sun, C.M., Vallone, M.K., Woo, T.M., Molineaux, C.J., and Bennett, M.K. (2007). Antitumor activity of PR-171, a novel irreversible inhibitor of the proteasome. Cancer Res 67, 6383–6391.

Dick, L.R., Cruikshank, A.A., Grenier, L., Melandri, F.D., Nunes, S.L., and Stein, R.L. (1996). Mechanistic studies on the inactivation of the proteasome by lactacystin: a central role for clasto-lactacystin beta-lactone. J Biol Chem 271, 7273–7276.

Dick, L.R., Cruikshank, A.A., Destree, A.T., Grenier, L., McCormack, T.A., Melandri, F.D., Nunes, S.L., Palombella, V.J., Parent, L.A., Plamondon, L., and Stein, R.L. (1997). Mechanistic studies on the inactivation of the proteasome by lactacystin in cultured cells. J Biol Chem 272, 182–188.

Elofsson, M., Splittgerber, U., Myung, J., Mohan, R., and Crews, C.M. (1999). Towards subunit-specific proteasome inhibitors: synthesis and evaluation of peptide alpha',beta'-epoxyketones. Chem Biol 6, 811–822.

Feling, R.H., Buchanan, G.O., Mincer, T.J., Kauffman, C.A., Jensen, P.R., and Fenical, W. (2003). Salinosporamide A: a highly cytotoxic proteasome inhibitor from a novel microbial source, a marine bacterium of the new genus salinospora. Angew Chem Int Ed Engl 42, 355–357.

Fenteany, G., Standaert, R.F., Lane, W.S., Choi, S., Corey, E.J., and Schreiber, S.L. (1995). Inhibition of proteasome activities and subunit-specific amino-terminal threonine modification by lactacystin. Science 268, 726–731.

Fevig, J.M., Buriak, J., Jr., Cacciola, J., Alexander, R.S., Kettner, C.A., Knabb, R.M., Pruitt, J.R., Weber, P.C., and Wexler, R.R. (1998). Rational design of boropeptide thrombin inhibitors: beta,

beta-dialkyl- phenethylglycine P2 analogs of DuP 714 with greater selectivity over complement factor I and an improved safety profile. Bioorg Med Chem Lett 8, 301–306.

Figueiredo-Pereira, M.E., Berg, K.A., and Wilk, S. (1994). A new inhibitor of the chymotrypsin-like activity of the multicatalytic proteinase complex (20S proteasome) induces accumulation of ubiquitin- protein conjugates in a neuronal cell. J Neurochem 63, 1578–1581.

Figueiredo-Pereira, M.E., Chen, W.E., Li, J., and Johdo, O. (1996). The antitumor drug aclacinomycin A, which inhibits the degradation of ubiquitinated proteins, shows selectivity for the chymotrypsin-like activity of the bovine pituitary 20S proteasome. J Biol Chem 271, 16455–16459.

Fitzpatrick, L.R., Khare, V., Small, J.S., and Koltun, W.A. (2006). Dextran sulfate sodium-induced colitis is associated with enhanced low molecular mass polypeptide 2 (LMP2) expression and is attenuated in LMP2 knockout mice. Dig Dis Sci 51, 1269–1276.

Fitzpatrick, L.R., Small, J.S., Poritz, L.S., McKenna, K.J., and Koltun, W.A. (2007). Enhanced intestinal expression of the proteasome subunit low molecular mass polypeptide 2 in patients with inflammatory bowel disease. Dis Colon Rectum 50, 337–348; discussion 348–350.

Groll, M., Ditzel, L., Lowe, J., Stock, D., Bochtler, M., Bartunik, H.D., and Huber, R. (1997). Structure of 20S proteasome from yeast at 2.4 A resolution. Nature 386, 463–471.

Groll, M., Kim, K.B., Kairies, N., Huber, R., and Crews, C.M. (2000). Crystal structure of epoxomicin:20S proteasome reveals a molecular basis for selectivity of α',β'-epoxyketone proteasome inhibitors. J Am Chem Soc 122, 1237–1238.

Groll, M., Koguchi, Y., Huber, R., and Kohno, J. (2001). Crystal structure of the 20 S proteasome: TMC-95A complex: a non-covalent proteasome inhibitor. J Mol Biol 311, 543–548.

Groll, M., Huber, R., and Potts, B.C. (2006). Crystal structures of Salinosporamide A (NPI-0052) and B (NPI-0047) in complex with the 20S proteasome reveal important consequences of beta-lactone ring opening and a mechanism for irreversible binding. J Am Chem Soc 128, 5136–5141.

Hanada, M., Sugawara, K., Kaneta, K., Toda, S., Nishiyama, Y., Tomita, K., Yamamoto, H., Konishi, M., and Oki, T. (1992). Epoxomicin, a new antitumor agent of microbial origin. J Antibiot (Tokyo) 45, 1746–1752.

Harvey, A.L. (1999). Medicines from nature: are natural products still relevant to drug discovery? Trends Pharmacol Sci 20, 196–198.

Hershko, A., and Ciechanover, A. (1998). The ubiquitin system. Annu Rev Biochem 67, 425–479.

Ho, Y.K., Bargagna-Mohan, P., Mohan, R., and Kim, K.B. (2007). LMP2-specific inhibitors: Novel chemical genetic tools for proteasome biology. Chem Biol 14, 419–430.

Joazeiro, C.A., Anderson, K.C., and Hunter, T. (2006). Proteasome inhibitor drugs on the rise. Cancer Res 66, 7840–7842.

Karin, M., and Ben-Neriah, Y. (2000). Phosphorylation meets ubiquitination: the control of NF-[kappa]B activity. Annu Rev Immunol 18, 621–663.

Kazi, A., Daniel, K.G., Smith, D.M., Kumar, N.B., and Dou, Q.P. (2003). Inhibition of the proteasome activity, a novel mechanism associated with the tumor cell apoptosis-inducing ability of genistein. Biochem Pharmacol 66, 965–976.

Kessler, B.M., Tortorella, D., Altun, M., Kisselev, A.F., Fiebiger, E., Hekking, B.G., Ploegh, H.L., and Overkleeft, H.S. (2001). Extended peptide-based inhibitors efficiently target the proteasome and reveal overlapping specificities of the catalytic beta-subunits. Chem Biol 8, 913–929.

Kim, K.B., and Crews, C.M. (2003). Natural product and synthetic proteasome inhibitors. In Cancer Drug Discovery and Development: Proteasome Inhibitors in Cancer Therapy, J. Adams, ed. (Totowa, NJ: Humana Press Inc.), pp. 47–63.

King, R.W., Deshaies, R.J., Peters, J.M., and Kirschner, M.W. (1996). How proteolysis drives the cell cycle. Science 274, 1652–1659.

Kisselev, A.F., and Goldberg, A.L. (2001). Proteasome inhibitors: from research tools to drug candidates. Chem Biol 8, 739–758.

Kisselev, A.F., Callard, A., and Goldberg, A.L. (2006). Importance of the different proteolytic sites of the proteasome and the efficacy of inhibitors varies with the protein substrate. J Biol Chem 281, 8582–8590.

Kloetzel, P.M. (2001). Antigen processing by the proteasome. Nat Rev Mol Cell Biol 2, 179–187.

Koguchi, Y., Kohno, J., Nishio, M., Takahashi, K., Okuda, T., Ohnuki, T., and Komatsubara, S. (2000). TMC-95A, B, C, and D, novel proteasome inhibitors produced by Apiospora montagnei Sacc. TC 1093. Taxonomy, production, isolation, and biological activities. J Antibiot (Tokyo) 53, 105–109.

Koguchi, Y., Kohno, J., Suzuki, S., Nishio, M., Takahashi, K., Ohnuki, T., and Komatsubara, S. (2000). TMC-86A, B and TMC-96, new proteasome inhibitors from Streptomyces sp. TC 1084 and Saccharothrix sp. TC 1094. II. Physico-chemical properties and structure determination. J Antibiot (Tokyo) 53, 63–65.

Kohno, J., Koguchi, Y., Nishio, M., Nakao, K., Kuroda, M., Shimizu, R., Ohnuki, T., and Komatsubara, S. (2000). Structures of TMC-95A-D: novel proteasome inhibitors from Apiospora montagnei sacc. TC 1093. J Org Chem 65, 990–995.

Loidl, G., Groll, M., Musiol, H.J., Huber, R., and Moroder, L. (1999a). Bivalency as a principle for proteasome inhibition. Proc Natl Acad Sci U S A 96, 5418–5422.

Loidl, G., Groll, M., Musiol, H.J., Ditzel, L., Huber, R., and Moroder, L. (1999b). Bifunctional inhibitors of the trypsin-like activity of eukaryotic proteasomes. Chem Biol 6, 197–204.

Loidl, G., Musiol, H.J., Groll, M., Huber, R., and Moroder, L. (2000). Synthesis of bivalent inhibitors of eucaryotic proteasomes. J Pept Sci 6, 36–46.

Lowe, J., Stock, D., Jap, B., Zwickl, P., Baumeister, W., and Huber, R. (1995). Crystal structure of the 20S proteasome from the archaeon T. acidophilum at 3.4 A resolution. Science 268, 533–539.

Lu, C., Gallegos, R., Li, P., Xia, C.Q., Pusalkar, S., Uttamsingh, V., Nix, D., Miwa, G.T., and Gan, L.S. (2006). Investigation of drug-drug interaction potential of bortezomib in vivo in female Sprague-Dawley rats and in vitro in human liver microsomes. Drug Metab Dispos 34, 702–708.

Meng, L., Kwok, B.H., Sin, N., and Crews, C.M. (1999a). Eponemycin exerts its antitumor effect through the inhibition of proteasome function. Cancer Res 59, 2798–2801.

Meng, L., Mohan, R., Kwok, B.H., Elofsson, M., Sin, N., and Crews, C.M. (1999b). Epoxomicin, a potent and selective proteasome inhibitor, exhibits in vivo antiinflammatory activity. Proc Natl Acad Sci U S A 96, 10403–10408.

Myung, J., Kim, K.B., and Crews, C.M. (2001a). The ubiquitin-proteasome pathway and proteasome inhibitors. Med Res Rev 21, 245–273.

Myung, J., Kim, K.B., Lindsten, K., Dantuma, N.P., and Crews, C.M. (2001b). Lack of proteasome active site allostery as revealed by subunit-specific inhibitors. Mol Cell 7, 411–420.

Nam, S., Smith, D.M., and Dou, Q.P. (2001). Ester bond-containing tea polyphenols potently inhibit proteasome activity in vitro and in vivo. J Biol Chem 276, 13322–13330.

Nazif, T., and Bogyo, M. (2001). Global analysis of proteasomal substrate specificity using positional-scanning libraries of covalent inhibitors. Proc Natl Acad Sci U S A 98, 2967–2972.

Oinonen, C., and Rouvinen, J. (2000). Structural comparison of Ntn-hydrolases. Protein Sci 9, 2329–2337.

Omura, S., Fujimoto, T., Otoguro, K., Matsuzaki, K., Moriguchi, R., Tanaka, H., and Sasaki, Y. (1991). Lactacystin, a novel microbial metabolite, induces neuritogenesis of 000neuroblastoma cells. J Antibiot (Tokyo) 44, 113–116.

Orlowski, M. (1993). The multicatalytic proteinase complex (proteasome) and intracellular protein degradation: diverse functions of an intracellular particle. J Lab Clin Med 121, 187–189.

Orlowski, R., and Orlowski, M. (2006). Potent and specific immunoproteasome inhibitors United States Patent Application 20060241056.

Orlowski, R.Z. (2005). The ubiquitin proteasome pathway from bench to bedside. Hematology (Am Soc Hematol Educ Program), 220–225.

Ostrowska, H., Wojcik, C., Omura, S., and Worowski, K. (1997). Lactacystin, a specific inhibitor of the proteasome, inhibits human platelet lysosomal cathepsin A-like enzyme. Biochem Biophys Res Commun 234, 729–732.

Ostrowska, H., Wojcik, C., Wilk, S., Omura, S., Kozlowski, L., Stoklosa, T., Worowski, K., and Radziwon, P. (2000). Separation of cathepsin A-like enzyme and the proteasome: evidence that lactacystin/beta-lactone is not a specific inhibitor of the proteasome. Int J Biochem Cell Biol 32, 747–757.

Palmer, J.T., Rasnick, D., Klaus, J.L., and Bromme, D. (1995). Vinyl sulfones as mechanism-based cysteine protease inhibitors. J Med Chem 38, 3193–3196.

Rock, K.L., Gramm, C., Rothstein, L., Clark, K., Stein, R., Dick, L., Hwang, D., and Goldberg, A.L. (1994). Inhibitors of the proteasome block the degradation of most cell proteins and the generation of peptides presented on MHC class I molecules. Cell 78, 761–771.

Ruiz, S., Krupnik, Y., Keating, M., Chandra, J., Palladino, M., and McConkey, D. (2006). The proteasome inhibitor NPI-0052 is a more effective inducer of apoptosis than bortezomib in lymphocytes from patients with chronic lymphocytic leukemia. Mol Cancer Ther 5, 1836–1843.

Sadaghiani, A.M., Verhelst, S.H., and Bogyo, M. (2007). Tagging and detection strategies for activity-based proteomics. Curr Opin Chem Biol 11, 20–28.

Sekizawa, R., Momose, I., Kinoshita, N., Naganawa, H., Hamada, M., Muraoka, Y., Iinuma, H., and Takeuchi, T. (2001). Isolation and structural determination of phepropeptins A, B, C, and D, new proteasome inhibitors, produced by Streptomyces sp. J Antibiot (Tokyo) 54, 874–881.

Sin, N., Meng, L., Auth, H., and Crews, C.M. (1998). Eponemycin analogues: syntheses and use as probes of angiogenesis. Bioorg Med Chem 6, 1209–1217.

Sin, N., Kim, K.B., Elofsson, M., Meng, L., Auth, H., Kwok, B.H., and Crews, C.M. (1999). Total synthesis of the potent proteasome inhibitor epoxomicin: a useful tool for understanding proteasome biology. Bioorg Med Chem Lett 9, 2283–2288.

Spaltenstein, A., Leban, J.J., Huang, J.J., Reinhardt, K.R., Viveros, O.H.S.J., and Crouch, R. (1996). Design and synthesis of novel protease inhibitors. Tripeptide α',β'-epoxyketones as nanomolar inactivators of proteasome. Tetrahedron Lett. 37, 1343–1346.

Sugawara, K., Hatori, M., Nishiyama, Y., Tomita, K., Kamei, H., Konishi, M., and Oki, T. (1990). Eponemycin, a new antibiotic active against B16 melanoma. I. Production, isolation, structure and biological activity. J Antibiot (Tokyo) 43, 8–18.

Tan, G., Gyllenhaal, C., and Soejarto, D.D. (2006). Biodiversity as a source of anticancer drugs. Curr Drug Targets 7, 265–277.

van Swieten, P.F., Samuel, E., Hernandez, R.O., van den Nieuwendijk, A.M., Leeuwenburgh, M.A., van der Marel, G.A., Kessler, B.M., Overkleeft, H.S., and Kisselev, A.F. (2007). A cell-permeable inhibitor and activity-based probe for the caspase-like activity of the proteasome. Bioorg Med Chem Lett 17, 3402–3405.

Verdoes, M., Berkers, C.R., Florea, B.I., van Swieten, P.F., Overkleeft, H.S., and Ovaa, H. (2006a). Chemical proteomics profiling of proteasome activity. Methods Mol Biol 328, 51–69.

Verdoes, M., Florea, B.I., Menendez-Benito, V., Maynard, C.J., Witte, M.D., van der Linden, W.A., van den Nieuwendijk, A.M., Hofmann, T., Berkers, C.R., van Leeuwen, F.W., Groothuis, T.A., Leeuwenburgh, M.A., Ovaa, H., Neefjes, J.J., Filippov, D.V., van der Marel, G.A., Dantuma, N.P., and Overkleeft, H.S. (2006b). A fluorescent broad-spectrum proteasome inhibitor for labeling proteasomes in vitro and in vivo. Chem Biol 13, 1217–1226.

Verdoes, M., Hillaert, U., Florea, B.I., Sae-Heng, M., Risseeuw, M.D., Filippov, D.V., van der Marel, G.A., and Overkleeft, H.S. (2007). Acetylene functionalized BODIPY dyes and their application in the synthesis of activity based proteasome probes. Bioorg Med Chem Lett 17, 6169–6171.

Verma, R., and Deshaies, R.J. (2000). A proteasome howdunit: the case of the missing signal. Cell 101, 341–344.

Wilk, S., and Figueiredo-Pereira, M.E. (1993). Synthetic inhibitors of the multicatalytic proteinase complex (proteasome). Enzyme Protein 47, 306–313.

Wilk, S., and Orlowski, M. (1983a). Evidence that pituitary cation-sensitive neutral endopeptidase is a multicatalytic protease complex. J Neurochem 40, 842–849.

Wilk, S., and Orlowski, M. (1983b). Inhibition of rabbit brain prolyl endopeptidase by n-benzyloxycarbonyl-prolyl-prolinal, a transition state aldehyde inhibitor. J Neurochem 41, 69–75.

Yang, H., Chen, D., Cui, Q.C., Yuan, X., and Dou, Q.P. (2006). Celastrol, a triterpene extracted from the Chinese "Thunder of God Vine," is a potent proteasome inhibitor and suppresses human prostate cancer growth in nude mice. Cancer Res 66, 4758–4765.

Yang, H., Shi, G., and Dou, Q.P. (2007). The tumor proteasome is a primary target for the natural anticancer compound Withaferin A isolated from "Indian winter cherry". Mol Pharmacol 71, 426–437.

Yewdell, J.W. (2005). Immunoproteasomes: regulating the regulator. Proc Natl Acad Sci U S A 102, 9089–9090.

Targeting Deubiquitinating Enzymes

Carmen Priolo, Derek Oldridge, Martin Renatus, and Massimo Loda

Abstract Deubiquitinating enzymes (DUBs) or isopeptidases belong to the enzyme class of hydrolases which include more than 100 members identified thus far. These proteins, which are grouped into four cysteine protease families and one metallo-protease family, catalyze the removal of ubiquitin from specific protein targets by cleavage of the linking isopeptide bond. DUB activity has been found to affect critical events in the cell proteome, including protein degradation through the proteasome machinery as well as endocytosis and lysosomal degradation of membrane receptors, with implications in cell signal transduction, cell cycle regulation, DNA damage tolerance, inflammatory response, and ribosomal protein synthesis. Strong evidence for an involvement of DUBs in the pathogenesis of numerous human diseases, including cancer, makes members of this class appealing potential targets for new anti-tumoral therapies, and has been nourishing a series of studies and chemical screens devoted to the discovery of compounds to interfere with DUB activity. However, much still remains unknown about the biology of isopeptidases, and extensive efforts of integrating crystallographic, mechanistic and functional data are needed to design new therapeutic strategies targeting the ubiquitin system.

Keywords Deubiquitinating enzymes • Isopeptidases • Ubiquitin Specific Proteases • Cancer • Therapy

Introduction

Ubiquitination is a post-translational modification of proteins through the covalent addition of Ubiquitin (Ub), a 76-residue polypeptide that is highly conserved among eukaryotes but absent in prokaryotes and archeae. Chemically, the carboxyl

C. Priolo, D. Oldridge, and M. Loda (✉)
Departments of Medical Oncology and Pathology, and the Center for Molecular Oncologic Pathology, Dana Farber Cancer Institute and Brigham and Women's Hospital, Harvard Medical School, Boston, MA, USA
e-mail: massimo_loda@dfci.harvard.edu

M. Renatus
Novartis Institutes for BioMedical Research, Basel, Switzerland

K. Sakamoto and E. Rubin (eds.), *Modulation of Protein Stability in Cancer Therapy*,
DOI: 10.1007/978-0-387-69147-3_8, © Springer Science+Business Media, LLC 2009

group of the C-terminal glycine of ubiquitin forms an isopeptide bond with the ε-aminogroup of a lysine residue of the target protein (Pickart and Eddins 2004).

Ubiquitin modification can occur by conjugation of an individual Ub molecule (monoubiquitination) or a poly-Ub chain (polyubiquitination) to a single lysine residue of the protein substrate, or through single Ub molecule linkage to several different lysine residues (multiubiquitination) of the protein substrate. Each of these modifications is thought to determine a different destiny for the protein. While monoubiquitination and multiubiquitination are involved in endocytosis and intracellular trafficking, polyubiquitination, long considered to be exclusively required for proteasomal degradation, might actually play a role in multiple additional cellular functions, such as DNA damage response, inflammation, endocytosis, and ribosomal protein synthesis (Hicke 2001; Kirkin and Dikic 2007; Mukhopadhyay and Riezman 2007). To this end, the specific lysine linkage in polyubiquitin chains has recently emerged as a functional signal in pathway activation. In fact, the ubiquitin molecule itself contains seven lysine residues that can mediate ubiquitin polymerization prior to protein conjugation and, among these seven internal lysines, linkages through Lys^{48} (Ub-Lys^{48}) are required for targeting the proteasome, whereas Ub-Lys^{63} chains are required for endocytosis and the other functions listed above (Pickart and Fushman 2004).

Conjugation of ubiquitin to its protein substrate requires an enzymatic cascade: first, a ubiquitin-activating enzyme (E1) forms an ATP-dependent thioester bond with ubiquitin, in which the terminal glycine of ubiquitin is linked to the thiol group of a cysteine residue in the E1 active site, and transfers the ubiquitin to a ubiquitin-conjugating enzyme (E2); next, E2 forms a thioester bond with the activated ubiquitin and transfers it to the target; finally, a ubiquitin-ligase (E3) catalyzes the isopeptide bond formation between ubiquitin and the specific protein substrate (Pickart and Eddins 2004).

Deubiquitinating enzymes (DUBs), or isopeptidases, catalyze the removal of ubiquitin from protein substrates, thus functionally affecting critical events in the cell proteome, such as protein degradation through the proteasome machinery, internalization, lysosomal degradation of membrane receptors, and ubiquitin recycling (Amerik and Hochstrasser 2004; Wilkinson 1997; Wing 2003). Since these cellular events have been shown to be important in human disease, including human tumorigenesis and tumor progression, DUBs have emerged as an appealing potential target for new anti-cancer therapies (Yang 2007). However, much still remains unknown about this class of enzymes.

In the last decade, a number of studies have been performed to identify new DUBs and explore their multiple functions in human disease, as well as to assay DUB activity in vitro to screen libraries of pharmacological molecules. Importantly, structures of several DUBs have been solved, thus providing crucial information for the discovery of drugs.

In this chapter, a classification of the known DUB family members will be reviewed by both chemical and functional approaches, and relevant examples of DUBs involved in tumorigenesis, as well as potential drug targeting strategies, will be discussed.

Classification and Biochemistry
of Deubiquitinating Enzymes (DUBs)

The understanding of the manifold roles of DUBs in homeostasis and disease is still in its infancy. The proteolytic activity of these enzymes can result in one or more of the following cellular effects: increased protein target half-life, modification of protein target subcellular localization, and increase of free ubiquitin intracellular concentration.

Phylogenetically, DUBs are present in proteobacteria and Chlamydiae and were probably originally transmitted to host cells by lateral gene transfer (Horn et al. 2004). In addition, the highly conserved Ub is thought to be the oldest and most important modifier in eukaryotes (Iyer et al. 2006).

DUBs belong to the enzyme class of hydrolases, which can be classified into five classes–aspartic, metallo, serine, threonine, and cysteine proteases–each with a different mechanism of catalysis (Puente and Lopez-Otin 2004). Based on sequence similarities of the Ub-protease domain and putative mechanisms of action, DUBs fall into at least five subclasses, four of which belong to the cysteine-protease family and one of which belongs to the metallo-protease family. Despite the conservation of mechanism, the secondary and tertiary structures of the four known cysteine protease families are highly diverse.

The catalytic activity of cysteine proteases mainly depends on the thiol group of a cysteine residue in the active site, which is activated (deprotonated) by an adjacent histidine, itself often polarized by an aspartate residue. Collectively, these three aminoacid residues form the catalytic triad. A nucleophilic attack of the peptide bond by the thiolate (activated cysteine) results in an charged tetragonal intermediate – which is stabilized by the oxyanion hole formed by residues such as an asparagines, glutamine or glutamate residue – and in the release of the target as well as the generation of a covalent intermediate with the Ub moiety. This is eventually released by hydrolysis through the addition of a water molecule. Conversely, metalloproteases use a Zn^{2+}-bound polarized water molecule to generate a noncovalent intermediate with the substrate, which is released by a proton transfer from a water molecule.

The following four subclasses of DUBs are cysteine proteases: ubiquitin-specific protease (USP), ubiquitin C-terminal hydrolase (UCH), Otubain protease (OTU), and Machado–Joseph disease protease (MJD). JAMM motif proteases are instead metalloproteases (Amerik and Hochstrasser 2004; Nijman et al. 2005; http://merops.sanger.ac.uk/).

Nijman et al. (2005) annotated 95 putative DUBs in the human genome by selecting genes whose transcripts encode one of the five protease domains using the ENSEMBL human genome database. In detail, 58 USP, 4 UCH, 5 MJD, 14 OUT, and 14 JAMM domain-containing genes, many of which associated with multiple transcripts, were found. Moreover, all but five predicted DUBs are expressed in humans, according to the NCBI human-expressed sequence tag (EST) databases (Nijman et al. 2005).

Ubiquitin-Specific Proteases (USPs)

The largest subclass, including most of the DUBs encoded by the human genome, is the USP family of proteases, which is characterized by two short and well-conserved motifs, named Cys and His boxes (Fig. 1). These two motifs include residues at the catalytic center crucial for substrate recognition and catalysis, namely the catalytic cysteine (Cys) and histidine (His). Among USPs, the size of the entire catalytic domain lies between 300 and 800 aminoacids; this is due to the variability of the unrelated sequences which interspace the two motifs and are potentially involved in a protein regulatory function (Nijman et al. 2005; Quesada et al. 2004). Proteases belonging to this family have been implicated in tumorigenesis (USP2, CYLD,

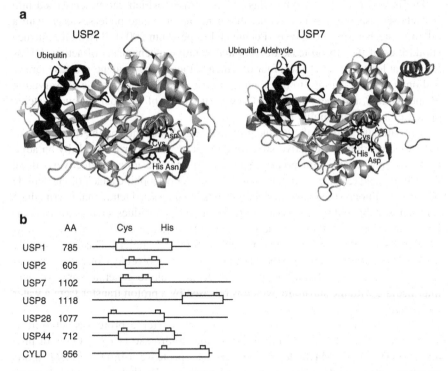

Fig. 1 (a) The overall crystal structures of USP2 (PDB entry 2HD5) and USP7 (PDB entry 1NBF) catalytic domains in complex with ubiquitin and ubiquitin aldehyde, respectively, show high similarity, each resembling a cupped hand with three subdomains, that are referred to as finger, palm, and thumb domains. The ubiquitin core rests within the pocket created by the cupped hand, whereas the C-terminal residues thread through a narrow channel leading toward the catalytic triad (Cys, His, Asp/Asn), which resides between the palm and thumb. Structures shown here were visualized using PyMOL (DeLano 2002). (b) Schematic sequence domains of representative USPs, characterized by two short and well-conserved motifs named Cys and His boxes. Among these isopeptidases, the size of the entire catalytic domain can change, due to the variability of unrelated sequences that interspace the two motifs. Higher variability is conferred by the length of the amino-terminal domain, which is thought to be involved in binding of selected substrates

USP7), receptor tyrosine kinase internalization/degradation (USP8), cell cycle regulation (USP44) and DNA damage response (USP1, USP28).

Ubiquitin C-Terminal Hydrolases (UCH)

The human UCH family of proteases was the first to be identified and consists of four human members. These enzymes were named for their activity in hydrolyzing small amides and esters at the carboxyl terminus of ubiquitin (Pickart and Rose 1985). Two members of this group seem to be involved in human disease: inactivating mutations of UCH-L1 have been described in neurodegenerative disease such as Parkinson's and Alzheimer's diseases (Setsuie and Wada 2007), and mRNA over-expression of both UCH-L1 and UCH-L3 has been associated with the biologically aggressive breast cancer (Miyoshi et al. 2006).

Ovarian Tumor-Related Proteases (OTUs)

This subclass of DUBs was predicted by a bioinformatic approach (Makarova et al. 2000) using the Drosophila Otu gene (involved in ovary development) and its homologs to find sequential similarities with genes encoding viral cysteine proteases. So far, three OTU proteins have been shown to display DUB activity in vitro: Otubain-1, Otubain-2 (both isolated from HeLa cervical cancer cells by affinity purification) (Balakirev et al. 2003) and Cezanne (Evans et al. 2003). In addition, the protein A20 contains both a DUB domain belonging to the OTU subfamily at the amino-terminus and a ligase domain at the carboxyl-terminus. This enzyme is involved in NF-kappa B signaling and its multiple functions need to be further explored (Wertz et al. 2004).

Machado–Joseph Disease Protease (MJD)

Ataxin-3 and Ataxin-3-like proteins are proteases identified as members of this subfamily. Of relevance, Ataxin-3 gene alteration caused by instability of a CAG nucle-otide repeat leads to an inherited neurological condition known as spinocerebellar ataxia type-3 or Machado–Joseph disease (Scheel et al. 2003), whose molecular mechanisms of pathogenesis are still unclear.

JAMM Motif Proteases

Among the 14 members that have been predicted to belong to this family by Njiman et al. (2005), seven proteins show at least one amino acid change in the conserved

Zn^{2+} ion-stabilizing residues, thus suggesting that they may not be functional proteases. Interestingly, three JAMM isopeptidases have been implicated in cancer: AMSH plays a role in EGFR internalization/degradation (McCullough et al. 2004); CSN5 has been proposed as an oncogenic isopeptidase in breast cancer (Adler et al. 2008); BRCC36 is a BRCA1/2 interacting protein aberrantly expressed in the majority of breast tumors (Dong et al. 2003), which might intervene in tumor cell response to ionizing radiation (Chen et al. 2006; Sobhian et al. 2007).

Chemistry-Based Assays in DUB Discovery

In addition to in silico approaches, a variety of chemical tools have been applied to promote DUB discovery, including functional profiling, structural characterization, mechanistic classification and development of pharmacological compounds (Evans and Cravatt 2006; Hemelaar et al. 2004; Love et al. 2007; Quesada et al. 2004). As an example, the activity-based protein profiling (ABPP) technique, in which chemical probes containing an epitope-tagged Ub (HAUb) with a C-terminal thiol-reactive group are directed at the enzyme active sites as suicide substrates, has been applied to profiling of deubiquitinating enzymes with actual biological activity in the EL4 mouse thymoma cell line (Borodovsky et al. 2002), 23 DUBs including ten polypeptides for which no enzymatic activity had been previously demonstrated were targeted by these probes. Most of these proteases with specificity for Ub belong to the USP and UCH families. These results validate the theory that integrative approaches of both bioinformatics and function-dependent methodologies are required to explore new classes of proteins with enzymatic functions. In fact, although many DUBs can be easily identified by sequential alignment with family members, others may need an activity-based profiling approach.

Moreover, kinetic assays utilizing a variety of Ub-based probes have been applied in high-throughput screening (HTS) for DUB inhibitors, where a certain success has been achieved with the synthesis of fluorogenic Ub substrates (Love et al. 2007; Hassiepen et al. 2007). As an example, a Ub variant created by fusion with the N-terminal NusA tag and the C-terminal alpha-NH(2)-tetramethylrhodamin-lysine (fluorophore) has been used for the development of a miniaturized, fluorescence-polarization-based HTS assay for UCH-L3 and USP2 inhibitors (Tirat et al. 2005).

Structure of Isopeptidases

Crystallographic data are available for representative members of all DUB subclasses (Ambroggio et al. 2004; Hu et al. 2002, 2005, 2006; Johnston et al. 1997, 1999; Renatus et al. 2006; Sheng et al. 2006; Tran et al. 2003). The structures around the active center resemble the well-described papain-like cysteine proteases, a family of proteases that includes well-known drug targets such as cathepsin K (Vasiljeva et al. 2007). Despite a pronounced sequence diversity, essential residues of the catalytic machinery such as the active site catalytic triad of Cys-His-Asp184 as well as the oxyanion hole residue Gln 89 from USPs, UCHs and cathepsins superimpose very well (Amerik and Hochstrasser 2004; Johnston et al. 1997), indicating that the catalytic mechanism is conserved. Structural work on USP-7 and UCH-L3 (Hu et al. 2002; Johnston et al. 1997; Misaghi et al. 2005), where both the structure of the free enzyme and a covalent protease/ubiquitin complex are available, revealed that both enzymes undergo major structural changes around the active center upon ubiquitin binding. In the case of USP7, where the ligand-free form appears to adopt an inactive conformation, ubiquitin binding realigns the catalytic machinery, while in the case of UCH-L3 only in the ubiquitin bound structure are the substrate recognition sites fully formed. Based on analogy, it can be hypothesized that ubiquitin-induced three-dimensional rearrangements leading to the catalytically competent conformation are a feature common to deubiquitinating enzymes. However, the mechanistic implications of these structural changes have not yet been fully explored. Substrate-induced activation of proteases is thought to be one strategy to obtain absolute selectivity, as discussed for other proteases such as factor D (Taylor et al. 1999). Indeed, DUBs (data are available for USP-2, -5 and UCH-L1, -L3 (Luchansky et al. 2006; Melandri et al. 1996; Renatus et al. 2006; Stein et al. 1995)) show a high selectivity for ubiquitin based substrates. In fact, while ubiquitin based substrates are cleaved very efficiently–UCH-L3 cleaves ubiquitin- amc (aminomethylcoumarin) at rates close to diffusion control–short peptidic substrates based on the C-terminal sequence of ubiquitin are only poorly recognized by these enzymes or are not substrates at all. These mechanistic properties are a possible liability for the discovery of potent and selective site-directed inhibitors of deubiquitinating enzymes. In contrast to cathepsins, where the active site is preformed, a hypothetical small molecule inhibitor of a DUB will have to induce the catalytically competent conformation with a fully formed substrate-binding site. The enzymatic data available for UCH-L1 and -L3 suggest that these structural rearrangements will come at a significant energetic cost, lowering the affinity of such a ligand. Whether or not these mechanistic properties will impede the discovery of small molecule based inhibitors of DUBs remains to be seen.

   Of interest are the structures of two members of the USP subfamily: HAUSP (USP7) (Hu et al. 2002) and USP2 (Renatus et al. 2006) (Fig. 1), both protagonists in the regulation of proteins involved in human tumorigenesis (Mdm2, Fatty Acid Synthase (FASN), and the oncosuppressor gene p53) (Graner et al. 2004; Stevenson et al. 2007). As described, the tertiary structures of USP7 and USP2 each resemble a cupped hand, and their three subdomains are fittingly referred to as finger, palm, and thumb domains. As inferred from USP/ubiquitin aldehyde and USP/ubiquitin co-crystals, the ubiquitin core residues (1–71) rest within the pocket created by the cupped hand of its corresponding USP, whose complementary shape is believed to underlie ubiquitin binding specificity. The ubiquitin C-terminal residues (72–76) thread through a narrow channel leading toward the catalytic triad, which resides between the palm and thumb.

Deubiquitinating Enzymes and Therapeutic Perspectives in Cancer

Despite the number of DUBs known to date, specific protein substrates have been characterized for only a few of them. While early studies of DUBs focused on their general functions (ubiquitin recycling from protein degradation products, single ubiquitin cleavage from poly-ubiquitin chains), in the last decade numerous theories have been emerging regarding the protein target-specificity of DUBs. Moreover, a number of studies directed to the discovery of DUB interactors, and complemented by approaches of DUB overexpression or silencing (RNAi) in normal and cancer cell lines, have suggested potential roles for different members of this protease family in cancer progression, either as oncoproteins or tumor suppressors (Table 1). However, the expression levels and cellular/subcellular localization of specific DUBs in

Table 1 Representative DUBs linked to cancer

DUB	Target(s)	Disease	Links to cancer	References
USP1	FANCD2, PCNA	Fanconi anemia	DNA damage response	(Füjiwara et al. 1998; Nijman et al. 2005; Huang et al. 2006; Oestergaard et al. 2007)
USP2	FASN, MDM2	Prostate	Resistance to apoptosis	(Graner et al. 2004; Priolo et al. 2006; Stevenson et al. 2007)
USP7	p53, MDM2	Lung	Regulation of p53 stability	(Li et al. 2002, 2004; Cummins et al. 2004)
USP8	EGFR, ESCRTs	?	Membrane receptor internalization and degradation	(Mizuno et al. 2005; Row et al. 2006; Alwan et al. 2007)
USP28	53BP1, Pbw7α	Colon, lung and breast	DNA damage response and cell cycle regulation	(Valero et al. 2001; Zhang et al. 2006a; Popov et al. 2007a)
CYLD	TRAF2, NEMO	Familial cylin-dromatosis	NP-κB and JNK signaling	(Bignell et al. 2000; Brummelkamp et al. 2003; Kovalenko et al. 2003; Trompouki et al. 2003)
UCH-Li/3	?	Breast cancer	Overexpressed in breast cancer	(Miyoshi et al. 2006)
AMSH	EGFR	?	Membrane receptor internalization and degradation	(McCullough et al. 2004; Ma et al. 2007)
CSN5	?	Breast	MYC regulation, DNA damage response, amplified/overex-pressed in breast cancer	(Adler et al. 2006, 2008)
BRCC36	?	Breast	DNA damage response	(Dong et al. 2003 ; Chen et al. 2006; Sobhian et al. 2007)

human tumors is still lacking and further investigation is needed in order to cross the bridge between functional in vitro studies and therapeutics. Some of the most intriguing data suggestive of an involvement of DUBs in cancer are discussed below.

Usp7

USP7 was originally identified as a binding target of the herpes virus E3 ligase ICP0, from which its alternative name, herpes-associated ubiquitin specific protease (HAUSP), is derived (Everett et al. 1997). The first implication of its role in human carcinogenesis came with the discovery that the tumor suppressor p53 is stabilized by USP7-mediated deubiquitination, resulting in cell growth arrest and apoptosis (Li et al. 2002). However, this model was later confounded by the finding that MDM2, the E3 ligase targeting p53 for proteasomal degradation, is itself stabilized by USP7-mediated deubiquitination, and that complete ablation of USP7 has a net stabilizing effect on p53 by antagonizing MDM2 (Cummins et al. 2004; Li et al. 2004). Additionally, in a recent study of neuroblastoma cell lines, impairment of USP7 deubiquitinating activity toward p53 results in multiubiquitination and mislocalization of the latter, giving rise to inactive, cytoplasmically sequestered p53 (Becker et al. 2007). Although the function of USP7-mediated stabilization of both p53 and its E3 ligase, MDM2, is not fully understood, several models have been proposed (Brooks and Gu 2004). Based on the destabilization of MDM2 that accompanies USP7 ablation and subsequent up-regulation of p53, it has been argued that USP7 is a potential therapeutic target in cancers exhibiting p53 mis-regulation, but a low frequency of p53 mutations, such as hematopoietic cancers (Cheon and Baek 2006).

USP7 expression has been assessed in human primary tumors. USP7 levels were found reduced in 45% of non-small cell lung cancers (NSCLC) by immunohisto-chemistry, and mutation of p53 or reduction in USP7 was correlated with poor prognosis in NSCLC adenocarcinomas (Masuya et al. 2006). In another study, USP7 expression was detected in a limited number of cases of cervical carcinomas (19%) but at much higher frequencies in cervical cancer derived cell lines (88%) as well as in keratinocytes immortalized with E6/E7, leading the authors to argue that USP7 may play a role in cell transformation (Rolen et al. 2006). These results and the lack of a clear consensus on whether USP7 possesses ultimately oncogenic or tumor-suppressive activity may simply reflect the complexity of USP7-mediated regulation of the p53/MDM2 axis.

In addition to playing a significant role in regulating the activity of p53 and MDM2, USP7 has been shown to inhibit the transcriptional activity of FOXO4, a member of the forkhead box O (FOXO) family of transcription factors that negatively regulate oncogenic phosphoinositide-3 kinase (PI3K) signaling (van der Horst et al. 2006). Also of note, MDM2 activity has been linked to histone ubiquitination and transcriptional repression, another possible mechanism of MDM2 oncogenicity (Minsky and Oren 2004), and it will be interesting to see if USP7 may regulate transcriptional activity through histone deubiquitination, which has already been shown in *Drosophila* (van der Knaap et al. 2005). A recent proteome-wide screen identified 36 proteins whose expression was altered through shRNA-mediated knockdown of USP7, including proteins with roles in DNA replication, apoptosis, and endosomal organization (Kessler et al. 2007).

In summary, these results underscore that USP7 may play diverse roles in biological pathways of cancer and a variety of approaches is required to fully characterize this protease.

Usp2

USP2 was characterized in rat systems as a gene encoding two DUBs, UBP-t1 (Ubp45) and UBP-t2 (Ubp69) (due to alternate splicing of 5′ exons) (Lin et al. 2000), which are respective orthologues of the mouse Usp2-45 and Usp2-69 (Gousseva and Baker 2003) and of the human USP2b and USP2a (Graner et al. 2004) isoforms. The catalytic core, common to both isoforms, contains the "Cys box" and the "His box," which are the highly conserved motifs of the USP family. Gousseva and Baker (2003) reported a comprehensive analysis of USP2 expression in mouse embryo and adult tissues by western blot, immunohistochemistry and northern blot. Both Usp2-69 (human USP2a) protein and mRNA were consistently expressed at high levels in adult heart, testis and skeletal muscle, whereas no perfect concordance of the two was found in other organs. Interestingly, USP2 was confirmed by immunohistochemistry to have the highest expression in these organs during all phases of embryonic development. Furthermore, in some cell types the isoform Usp2-69 was typically localized in a perinuclear area and near the plasma membrane. In contrast, because of discrepancy between mRNA and protein expression, no clear information was available on Usp2-45 (human USP2b) abundance and subcellular localization. In keeping with the expression of USP2 in mouse embryonic muscle tissues, the rat USP2 isoforms have been shown to play a role in myoblast differentiation (Park et al. 2002), and one could therefore speculate an involvement of these proteins in leio- and/or rhabdomyosarcomas even though no data to support this theory are available yet.

Intriguingly, USP2, together with a few more genes, was identified as a target of circadian regulation in both mouse liver and heart tissues, suggesting that it could contribute to the core functions of all peripheral clocks – e.g., synchronization to hormone circulating factors, generation of rhythmicity and control of transcriptional outputs – by its protein stability function (Storch et al. 2002). A circadian expression of USP2 mRNA was also found in the suprachiasmatic nucleus (SCN), the master circadian pacemaker in mammals, and in both liver and SCN it seemed to be modulated by Clock, a basic helix-loop-helix (bHLH)-PAS transcription factor representing one of the main regulators of the circadian-dependent transcription in peripheral tissues (Oishi et al. 2003). Contrary to many other circadian genes, the rhythmic expression of USP2 was not regulated by glucocorticoids (Oishi et al. 2005). Both the circadian periodicity and the lack of dependency on adrenal hormone regulation need to be taken into account when devising therapeutic strategies against this isopeptidase.

Interestingly, among the genes that were co-regulated in a circadian fashion by Clock in mouse liver, one encodes for a USP2 substrate, namely Fatty acid synthase (FASN) (Oishi et al. 2003). Our group had recently demonstrated that the human isoform USP2a binds to and stabilizes FASN (Graner et al. 2004), which represents a key enzyme in lipid metabolism, responsible for the synthesis of palmitate. FASN is minimally expressed in most normal human tissues with the exception of the liver and adipose tissue (Kusakabe et al. 2000), whereas its expression markedly increases in several human cancers, being often associated with a poor prognosis (Alo et al. 2007; Rossi et al. 2003; Sebastiani et al. 2004; Visca et al. 2004; Zhao et al. 2006). Moreover, recent evidence points to FASN as a metabolic oncogene in breast and prostate cancer (Knowles and Smith 2007; Kuhajda 2006; Swinnen et al. 2006) suggesting that targeting USP2a, which prolongs its half-life, may be of value in these hormonally-regulated cancers.

Remarkably, almost half of the prostate tumors overexpress USP2a when compared to an adjacent normal tissue and these tumors display a characteristic gene expression signature. Indeed, a Gene Set Enrichment Analysis (GSEA) was performed on microarray data comparing two groups of prostate cancers expressing high and low USP2a, respectively, revealing that high USP2a-tumors are strongly associated with gene sets involved in both synthesis and metabolism of fatty acids, whereas only the "cell death" gene pathway and a gene set of pro-apoptotic p53 target genes (including *apaf-1, BID, BAX, p21, caspases 1, 6* and *9*) are significantly linked with those tumors expressing low USP2a levels (Priolo et al. 2006). While we previously described that FASN is a substrate of USP2a, Stevenson et al. (2007) have shown that USP2a binds to and deubiquitinates the ubiquitin ligase Mdm2, resulting in enhanced p53 degradation. This could contribute to a hypothetical anti-apoptotic function of USP2a, as substantiated by the gene profiling data in human tumors.

Interestingly, gene sets related to receptor tyrosine kinase (RTK) family members and those controlling endocytic trafficking, such as *ErbB3* and *rab* (Hoekstra et al. 2004; Szymkiewicz et al. 2004), were overexpressed in tumors with high USP2a levels as well (Priolo et al. 2006). Such results could tie into previous evidence of a peculiar perinuclear and sub-membrane subcellular localization of this protease (Gousseva and Baker 2003), as well as with our finding of its binding to clathrin heavy chain in affinity chromatography experiments (unpublished), suggesting a role of USP2a in RTK internalization pathways.

We have shown that USP2a is oncogenic as determined by canonical experiments of overexpression in nontransformed cells and knockdown in tumor cell lines. In fact, immortalized human prostate epithelial cells (PrEC) and mouse fibroblasts (NIH3T3) were transformed by forced expression of the enzyme, and a number of human cancer cell lines derived from prostate, breast, sarcoma and colon tumors underwent apoptosis following USP2a silencing. In the latter case, destabilization of both USP2a substrates, FASN and Mdm2, could be invoked as the cause of apoptosis. However, the relative role played by each of these proteins, when suppressed, is currently not known and may vary in different cell lines according to FASN expression levels and p53 gene status.

Ectopic USP2a expression conferred resistance to both Taxol and Cysplatin-induced apoptosis to immortalized human prostate epithelial cells (PrEC) (Priolo et al. 2006), suggesting that increasing USP2a levels allow tumor cells to escape chemotherapy-induced programmed cell death. The emerging role of taxanes-based drug treatments in the clinical management of prostate cancer (Cabrespine et al. 2006; Febbo et al. 2005) highlights therefore the importance of being able to define subsets of patients potentially refractory to this therapy. The immediate implication is that human tumors overexpressing USP2a could be identified as resistant to chemotherapeutic agents.

Taken together, these findings strongly support the involvement of USP2 in human tumorigenesis by controlling crucial networks that affect the apoptotic machinery and suggest this protease as a therapeutic target in prostate cancer.

Cyld

CYLD was originally identified as a tumor suppressor gene for familial cylindromatosis, an autosomal dominant predisposition to develop multiple tumors from skin append-ages (Bignell et al. 2000). However, CYLD deletion or inactivation may underlie all of familial cylindromatosis, multiple familial trichoepithelioma, and Brooke-Spiegler syndromes – initially identified as separate disorders characterized by skin neoplasms – which may be different phenotypic manifestations of the same genetic defect (Young et al. 2006). From combined high density SNP analysis and gene expression profiling of multiple myeloma primary tumors, it has been argued that deletion of CYLD, which resides on 16q12.1, may also explain poor prognosis associated with 16q LOH in these patients (Jenner et al. 2007). In addition to this, there is the decreased transcription and translation of CYLD in human colon and hepatocellular tumor cell lines and primary carcinomas as well (Hellerbrand et al. 2007).

Several recent studies have investigated the consequences of CYLD loss in murine models. Importantly, CYLD knock-out mice show no overt phenotype since they develop normally, have a normal life span, and do not spontaneously develop tumors (Massoumi et al. 2006; Reiley et al. 2006; Zhang et al. 2006b). Importantly, however, CYLD knock-out mice showed high susceptibility to chemically induced tumors, which enhance proliferation upon further chemical or UV light stimulation (Massoumi et al. 2006). In a similar study, CYLD knock-out mice developed more severe colonic inflammation and exhibited greater induction of colonic tumors relative to their littermate controls in a colitis-associated cancer model (Zhang et al. 2006b). Together, these studies underscore the important role CYLD plays in conferring resistance to tumor development and proliferation under tumor-stimulating conditions.

A connection between CYLD and the NF-kappaB pathway was demonstrated when it was found that CYLD promotes deubiquitination of K63-linked polyubiquitin chains from the tumor necrosis factor (TNF) receptor associated factor 2 (TRAF2) and from the NF-kappaB essential modulator (NEMO) (Brummelkamp et al. 2003;

Kovalenko et al. 2003; Trompouki et al. 2003). Ubiquitinated TRAF2 and NEMO promote formation of the IkappaB kinase (IKK) complex – of which NEMO is an essential subunit – that, in turn, promotes the phosphorylation and subsequent degradation of its substrate IkappaB, allowing NF-kappaB nuclear translocation and the activation of generally oncogenic NF-kappaB dependent pathways. By demonstrating that CYLD deubiquitinates and therefore deactivates TRAF2 and NEMO, these studies indicated that a plausible mechanism for CYLD-mediated tumor suppression may be through attenuation of NF-kappaB activity.

More recently, several studies have indicated that CYLD may act as a negative regulator of the c-Jun N-terminal kinase (JNK) pathway, which, independently or in tandem with NF-kappaB downregulation, may also account for the tumor suppressing function of CYLD (Reiley et al. 2004; Zhang et al. 2006b). Furthermore, JNK-mediated apoptosis is impaired by CYLD knock-out in Drosophila, resulting in an abnormal development (Xue et al. 2007). In the last few years, a series of studies have identified a plethora of putative CYLD substrates, including Bcl3, another NF-kappaB activator (Massoumi et al. 2006); Lck, a stimulator of T-cell development (Reiley et al. 2006); and RIPl, a stimulator of spermatogenesis (Wright et al. 2007). Undoubtedly, CYLD has emerged as a quite versatile regulator, and the sheer number of putative substrates identified to date suggests that CYLD may play important roles, which we are only beginning to understand, in many diverse biological pathways, including cancer.

DUBs and Endocytosis

Mono- and multi-ubiquitination, as well as polyubiquitination through Ub-Lys[63] chains, serve as a sorting mechanism for several endosomal cargo proteins, such as receptor tyrosine kinases (RTKs), MHC class I molecules and transferrin receptor, into the lumen of multivesicular bodies (MVB), which can directly fuse with lysosomes. This process, mediated by the interaction with an intricate protein network including the endosomal sorting complexes required for transport (ESCRTs), eventually results in membrane receptor down-regulation and degradation by lysosomal acid hydrolases. Alternatively, receptors that are endocytosed but not ubiquitinated (i.e., transferrin receptor) are recycled back to the plasma membrane from the early endosomes. (Clague and Urbe 2006; Duncan et al. 2006).

Endosomal sorting mechanisms of the epidermal growth factor receptor (EGFR) are among the most documented in the literature. Two DUBs have been described to affect lysosomal degradation of EGFR, with potential relevance in tumors: USP8 (also called UBPY) and AMSH. Despite their distinct catalytic domains, these proteases share several characteristic features; in fact, both proteins bind to the SH3 domain of STAM, a component of ESCRT-0 (Kaneko et al. 2003; Kato et al. 2000), and strongly colocalize with endosomes when their catalytic site is made inactive by a single point mutation (McCullough et al. 2004; Mizuno et al. 2005).

The main difference between the two proteases is that AMSH acts through disassembly of K63-linked polyubiquitin chains, whereas UBPY is able to process

K48- as well as K63-linked polyubiquitin chains to lower denomination forms in vitro (Clague and Urbe 2006). However, both proteases deubiquitinate monoubiquitinated RTKs (McCullough et al. 2006; Mizuno et al. 2005) and, intriguingly, controversial effects on EGFR down-regulation following AMSH and USP8 activity suppression have been reported. In fact, while some have argued that UBPY protease may act by preventing EGFR lysosomal degradation (Mizuno et al. 2005), others have shown that its knockdown delays degradation of EGFR and MET receptor (Alwan and van Leeuwen 2007; Row et al. 2006) (another RTK overexpressed in most human cancers (Christensen et al. 2005; Peruzzi and Bottaro 2006)); similarly, two recent studies have described either enhancement (McCullough et al. 2004) or inhibition (Ma et al. 2007) of EGFR degradation following AMSH silencing. Therefore, it is still debated whether a role of oncogene or oncosuppressor protein can be attributed to USP8 and AMSH.

In agreement with a hypothetical oncosuppressor role of USP8, its overexpression was originally shown to reduce foci formation in U2OS human osteosarcoma cells, while silencing resulted in an increase of proliferation in this cell line (Naviglio et al. 1998). In the same report, which provided evidence of the actual deubiquitinating activity of this protease and showed endogenous expression of USP8 in different human cancer cell lines, USP8 was described to behave in normal cells as a growth-regulated USP, that typically accumulates upon growth stimulation of starved human fibroblasts and decreases in response to growth arrest induced by cell-cell contact (Naviglio et al. 1998).

Targeted deletion of USP8 *in vivo* (Niendorf et al. 2007) results in embryonic lethality, and induced inactivation in adulthood causes fatal liver failure. Of relevance, these mice show decreased protein but not mRNA levels of different RTKs, including EGFR, c-met and ERBB3, coupled with an impairment of MAPK signaling (Niendorf et al. 2007). In addition, USP8 has been shown to have also an indirect role in the regulation of ERBB3 levels, by deubiquitinating and stabilizing its ubiquitin ligase Nrdp1 (Cao et al. 2007; Wu et al. 2004).

Importantly, there is consistent evidence that inhibition of USP8 activity, both *in vitro* and *in vivo*, results in an increase of global levels of ubiquitinated proteins and their concomitant accumulation on morphologically aberrant endosomes (Mizuno et al. 2006). Indeed, multiple proteins that belong to ESCRT complexes and are involved in the sorting machinery are ubiquitinated, and can be stabilized by USP8 (Mizuno et al. 2006; Polo et al. 2002; van Delft et al. 1997), whose general activity may therefore affect receptor sorting and degradation in an independent fashion.

DUBs and Response to DNA Damage

Usp1

USP1 was discovered during the sequencing of the human genome and was classified as an ubiquitin specific protease by sequence homology with other members of the

USP family and subsequent functional characterization (Fujiwara et al. 1998). Although it was originally suggested that USP1 might possess oncogenic or tumor-suppressive activity based on its location within a chromosomal region implicated in cancer – the short arm of chromosome 1 – this prediction was confirmed only recently, when USP1 was implicated in Fanconi anemia (FA), a rare hereditary disorder arising from impairment of DNA damage repair mechanisms and characterized by bone-marrow failure and predisposition to several malignancies, including acute myeloid leukemia (AML). A recent study showed that USP1 activity regulates an important component of the FA pathway, the FANCD2 (Fanconi anemia, complementation group D2) protein, which in its monoubiquitinated form is recruited to DNA damage foci along with other DNA repair proteins (Nijman et al. 2005). Indeed, USP1 ablation conferred greater resistance to mitomycin C-induced chromosomal breakage concurrent with stabilization of monoubiquitinated FANCD2. By deubiquitinating and therefore inhibiting FANCD2 signaling, it was proposed that hyperactivity of USP1 might therefore inhibit DNA damage repair, leading to the accumulation of chromosomal aberrations, which could promote tumorigenesis.

Interestingly, proliferating cell nuclear antigen (PCNA), a monoubiquitinated protein that promotes translation synthesis (TLS) in response to DNA damage, is also negatively regulated by USP1 (Huang et al. 2006). USP1 knockdown actually resulted in a higher mutation frequency in response to UV irradiation. This finding has obvious therapeutic implications in cancers with hyperactivation of PCNA and of error-prone TLS. Most recently, in a study undertaken in the DT40 chicken cell line, USP1 knockout actually caused increased sensitivity to DNA cross-linking agents, which the authors argued might result from titration of FANCD2 away from DNA damage foci, principally mediated by constitutively activated FANCD2 (Oestergaard et al. 2007). These collective results underscore that the altered DNA damage response in cancer could be manipulated by targeting isopeptidases.

Usp28

USP28, originally discovered by homology with USP25 and by functional characterization (Valero et al. 2001), has only recently emerged as a USP of interest in cancer. USP28 stabilizes several cell cycle checkpoint proteins involved in DNA damage response, including p53-binding protein 1 (53BP1), with which USP28 forms a complex (Zhang et al. 2006a). In the absence of USP28, these checkpoint proteins are degraded, attenuating DNA damage signals and abrogating p53 apoptotic function. Based on these findings, it was argued that USP28 may act as a tumor suppressor, functioning through pathways which mediate DNA damage-induced apoptosis.

In a subsequent independent study, USP28 was found to deubiquitinate and stabilize MYC oncoprotein through binding and antagonizing the ubiquitinating activity of the MYC-targeting E3 ligase, Fbw7alpha (Popov et al. 2007a). Although this model predicts USP28 to be a putative oncoprotein in apparent contrast to the study mentioned earlier, MYC turnover is likely involved in cellular DNA damage response, providing further evidence of an important role for USP28 in

DNA damage signaling (Popov et al. 2007b). Furthermore, in cell lines derived from breast, lung, and colon carcinomas as well as glioblastoma, shRNA-mediated knockdown of USP28 inhibited tumor cell growth while USP28 protein shows high nuclear staining in primary colon and breast adenocarcinomas (Popov et al. 2007a). Even though further experiments are warranted to clarify the precise role of this isopeptidase in the regulation of the DNA damage response, it is clear that USP28 plays an important role in tumorigenesis.

Cell Cycle Regulation by the Ubiquitin System

The eukaryotic cell cycle consists of articulated events whose ordinate succession is guaranteed by specific checkpoint complexes in such a way that cells, for instance, do not undergo chromosome segregation and mitosis before DNA replication is complete. Activation and inactivation of serine/threonine protein kinases called cyclin-dependent kinases (CDKs) are primarily responsible for the overall control of the progression of the cell cycle and the rapid response to stimuli such as growth factors and DNA damage. CDKs are activated by the association to cyclins and inactivated through their association to CDK inhibitors. They themselves are subject to rapid and specific regulation by proteolysis through the ubiquitin-proteasome system. Indeed, hyperactivation of positive cell cycle regulators and/or inactivation of negative regulators can favor tumor development (DeSalle and Pagano 2001; Santamaria and Ortega 2006; Sherr 1996). The ubiquitin-mediated proteolysis of different components of the cell cycle machinery is controlled by two major classes of ubiquitin ligases: the SCF (Skp-Cullin-F-box protein and ROC/Rbx/SAG RING finger protein) and APC/C (Anaphase Promoting Complex/Cyclosome) complexes, involved in the regulation of G1 to S phase transition and in the sister chromatid separation/exit from the mitosis steps, respectively. While a number of SCF ligases and relative substrates have already been identified in humans (DeSalle and Pagano 2001; Jin et al. 2004), the role undoubtedly played by isopeptidases in cell cycle regulation remains elusive. Complex mechanisms control degradation and stability of different proteins involved in cell proliferation that show growth arresting or pro-proliferative function, and one could choose to target either ubiquitination or deubiquitination of these proteins, in order to affect tumor growth. A case in point is represented by the CDK inhibitor p27, specifically targeted for proteasomal degradation in tumor cells (Loda et al. 1997), and β-catenin, a downstream signaling factor of the Wnt proliferation pathway (Polakis 2000), both substrates of the same SCF complex that differs only by the substrate-recognizing F-box protein. Inhibition of ubiquitin ligases would be ideal in the first case, whereas an enhanced degradation activity would be required in the case of β-catenin (Rolfe et al. 1997; Su et al. 2003; Sun 2003). The converse situation would apply to putative regulatory isopeptidases regulating this important pathway.

In addition, a major regulatory role of the ubiquitination-deubiquitination balance seems to be played in metaphase to anaphase mitotic transition. In fact,

USP44 has been characterized as a DUB involved in the regulation of APC complex. Its activity is required to prevent the premature activation of the APC complex by stabilizing the APC inhibitory Mad2-Cdc20 complex. USP44 is elevated in checkpoint-arrested mitotic cells, while rapidly degraded as cells exit from mitosis (Stegmeier et al. 2007). Further exploration of the mechanisms underlying these events will shed light on new strategies of targeting cell cycle core components in tumors.

These examples suggest that diverse targeting opportunities are possible if the cell cycle machinery in cancer is to be targeted. This can be achieved through selected targeting of proteins that either enhance ubiquitination or stabilize cell cycle regulators via pre-proteosomal deubiquitination.

Ubiquitin-Like Modifiers

Since the original discovery of ubiquitin, a class of ubiquitin-like (Ubl) small proteins has been defined and is emerging as a significant parallel mechanism of post-translational protein regulation. While each individual Ubl may exhibit low sequence homology to ubiquitin and to one another, the defining characteristics of this class of proteins include the conserved three dimensional structure – the β-grasp fold – of ubiquitin and the ability to conjugate to other proteins (Kerscher et al. 2006). Many of the molecular mechanisms required for ubiquitin processing and ligation to protein substrates are also conserved in Ubls, but each Ubl may require its own set of dedicated enzymes. For example, Ubl ligation proceeds by a pathway analogous to the ubiquitin E1/E2/E3 ligase pathway, and in some cases unique E1-like, E2-like, or E3-like ligases exist for each Ubl (Johnson et al. 1997; Johnson and Blobel 1997; Liakopoulos et al. 1998; Osaka et al. 1998). Furthermore, DUB homologs, termed Ubl-specific proteases (ULPs), have been found for several Ubls and presumably play similar roles to DUBs in regulating the biological pathways associated with their protein substrates (Li and Hochstrasser 2000; Schwienhorst et al. 2000).

Like their counterpart ubiquitin, Ubls have been implicated in a wide variety of pathways, including transcription, DNA repair, signal transduction, autophagy, and cell cycle control (Kerscher et al. 2006). Additionally, misregulation of one of the better studied Ubl modifications, sumoylation – the covalent addition of the Ubl SUMO to target proteins – is associated with several human diseases, including Huntington's and Alzheimer's (Neve 2003; Ross and Pickart 2004; Steffan et al. 2004). In acute promyelocytic leukemia (APL), a fusion between PML protein and retinoic acid receptor alpha impairs PML protein sumoylation, in turn disrupting the formation of nuclear PML bodies that possess tumor suppressing function (Zhong et al. 2000). Although Ubls are only discussed in brief here, further study of the Ubl protein family will provide deeper insights into the mechanisms and functions of ubiquitin-like pathways, their implications in human disease, and may inform the development of future therapies.

Concluding Remarks

In the last few years the clear evidence that DUBs are implicated in cancer has been nourishing a series of studies and chemical screens, devoted to the discovery of new compounds and small molecules to interfere with DUB activity. Three therapeutic scenarios could take place (Fig. 2): (1) a DUB promotes tumorigenesis

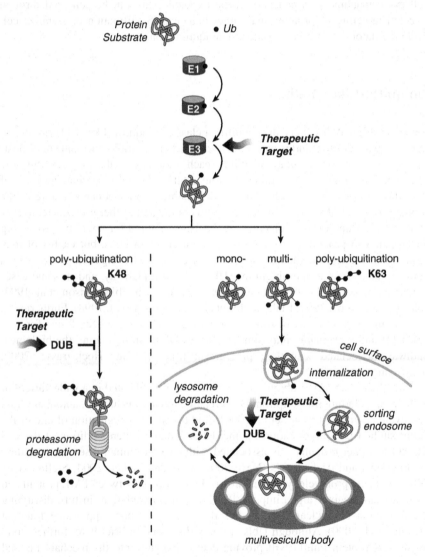

Fig. 2 Schematic representation of the ubiquitin system. *Thick arrows* point out potential therapeutic targets in cancer

(e.g., USP2) and molecules with inhibitory activity are sought after; (2) an isopeptidase exerts an oncosuppressor function (e.g., CYLD) and DUB activity needs to be restored to prevent disease; or (3) ubiquitin ligases (E3) themselves might constitute ideal therapeutic targets to either enhance or inhibit ubiquitination and degradation of specific protein substrates, especially if DUB targeting is otherwise intractable.

It should be emphasized that pharmacological approaches targeting DUBs and/or E3s are challenging, and extensive efforts of integrating crystallographic, mechanistic and functional data are needed to design and test therapeutic molecules. Unraveling the complex biology of DUBs (and, in general, of ubiquitin-related pathways) to select good targeted drugs will require a multifaceted approach that includes profiling DUBs or ULPs in tumors, possibly through activity-based assays, as also to assess the relevance of new proteases in cancer.

Acknowledgments This work was supported by grants from the NIH (RO1 CA131945, Specialized Programs of Research Excellence 5P50CA90381, PO1 CA089021), and the Prostate Cancer Foundation (M. Loda). M. Loda is a consultant for Novartis Pharmaceuticals, Inc., Dana Farber Cancer Institute.

References

Adler AS, Lin M, Horlings H, Nuyten DS, van de Vijver MJ, Chang HY (2006). Genetic regulators of large-scale transcriptional signatures in cancer Nat Genet. 38, 421-30

Adler, A. S., Littlepage, L. E., Lin, M., Kawahara, T. L., Wong, D. J., Werb, Z., and Chang, H. Y. (2008). CSN5 isopeptidase activity links COP9 signalosome activation to breast cancer progression. Cancer Res *68*, 506–515.

Alo, P. L., Amini, M., Piro, F., Pizzuti, L., Sebastiani, V., Botti, C., Murari, R., Zotti, G., and Di Tondo, U. (2007). Immunohistochemical expression and prognostic significance of fatty acid synthase in pancreatic carcinoma. Anticancer Res *27*, 2523–2527.

Alwan, H. A., and van Leeuwen, J. E. (2007). UBPY-mediated epidermal growth factor receptor (EGFR) de-ubiquitination promotes EGFR degradation. J Biol Chem *282*, 1658–1669.

Ambroggio, X. I., Rees, D. C., and Deshaies, R. J. (2004). JAMM: a metalloprotease-like zinc site in the proteasome and signalosome. PLoS Biol *2*, E2.

Amerik, A. Y., and Hochstrasser, M. (2004). Mechanism and function of deubiquitinating enzymes. Biochim Biophys Acta *1695*, 189–207.

Anal Biochem. 2007 Dec 15;371(2):201-7. A sensitive fluorescence intensity assay for deubiquitinating proteases using ubiquitin-rhodamine110-glycine as substrate.Hassiepen U, Eidhoff U, Meder G, Bulber JF, Hein A, Bodendorf U, Lorthiois E, Martoglio B.

Balakirev, M. Y., Tcherniuk, S. O., Jaquinod, M., and Chroboczek, J. (2003). Otubains: a new family of cysteine proteases in the ubiquitin pathway. EMBO Rep *4*, 517–522.

Becker, K., Marchenko, N. D., Maurice, M., and Moll, U. M. (2007). Hyperubiquitylation of wild-type p53 contributes to cytoplasmic sequestration in neuroblastoma. Cell Death Differ *14*, 1350–1360.

Bignell, G. R., Warren, W., Seal, S., Takahashi, M., Rapley, E., Barfoot, R., Green, H., Brown, C., Biggs, P. J., Lakhani, S. R., et al. (2000). Identification of the familial cylindromatosis tumour-suppressor gene. Nat Genet *25*, 160–165.

Borodovsky, A., Ovaa, H., Kolli, N., Gan-Erdene, T., Wilkinson, K. D., Ploegh, H. L., and Kessler, B. M. (2002). Chemistry-based functional proteomics reveals novel members of the deubiquitinating enzyme family. Chem Biol *9*, 1149–1159.

Brooks, C. L., and Gu, W. (2004). Dynamics in the p53-Mdm2 ubiquitination pathway. Cell Cycle *3*, 895–899.

Brummelkamp, T. R., Nijman, S. M., Dirac, A. M., and Bernards, R. (2003). Loss of the cylindromatosis tumour suppressor inhibits apoptosis by activating NF-kappaB. Nature *424*, 797–801.

Cabrespine, A., Guy, L., Khenifar, E., Cure, H., Fleury, J., Penault-Llorca, F., Kwiatkowski, F., Barthomeuf, C., Chollet, P., and Bay, J. O. (2006). Randomized Phase II study comparing paclitaxel and carboplatin versus mitoxantrone in patients with hormone-refractory prostate cancer. Urology *67*, 354–359.

Cao, Z., Wu, X., Yen, L., Sweeney, C., and Carraway, K. L., 3rd (2007). Neuregulin-induced ErbB3 downregulation is mediated by a protein stability cascade involving the E3 ubiquitin ligase Nrdp1. Mol Cell Biol *27*, 2180–2188.

Chen, X., Arciero, C. A., Wang, C., Broccoli, D., and Godwin, A. K. (2006). BRCC36 is essential for ionizing radiation-induced BRCA1 phosphorylation and nuclear foci formation. Cancer Res *66*, 5039–5046.

Cheon, K. W., and Baek, K. H. (2006). HAUSP as a therapeutic target for hematopoietic tumors (review). Int J Oncol *28*, 1209–1215.

Christensen, J. G., Burrows, J., and Salgia, R. (2005). c-Met as a target for human cancer and characterization of inhibitors for therapeutic intervention. Cancer Lett *225*, 1–26.

Clague, M. J., and Urbe, S. (2006). Endocytosis: the DUB version. Trends Cell Biol *16*, 551–559.

Cummins, J. M., Rago, C., Kohli, M., Kinzler, K. W., Lengauer, C., and Vogelstein, B. (2004). Tumour suppression: disruption of HAUSP gene stabilizes p53. Nature 428, 1 p following 486.

DeLano, W. L. (2002). The PyMOL Molecular Graphics System (Palo Alto, CA, USA: DeLano Scientific).

DeSalle, L. M., and Pagano, M. (2001). Regulation of the G1 to S transition by the ubiquitin pathway. FEBS Lett *490*, 179–189.

Dong, Y., Hakimi, M. A., Chen, X., Kumaraswamy, E., Cooch, N. S., Godwin, A. K., and Shiekhattar, R. (2003). Regulation of BRCC, a holoenzyme complex containing BRCA1 and BRCA2, by a signalosome-like subunit and its role in DNA repair. Mol Cell *12*, 1087–1099.

Duncan, L. M., Piper, S., Dodd, R. B., Saville, M. K., Sanderson, C. M., Luzio, J. P., and Lehner, P. J. (2006). Lysine-63-linked ubiquitination is required for endolysosomal degradation of class I molecules. Embo J *25*, 1635–1645.

Evans, M. J., and Cravatt, B. F. (2006). Mechanism-based profiling of enzyme families. Chem Rev *106*, 3279–3301.

Evans, P. C., Smith, T. S., Lai, M. J., Williams, M. G., Burke, D. F., Heyninck, K., Kreike, M. M., Beyaert, R., Blundell, T. L., and Kilshaw, P. J. (2003). A novel type of deubiquitinating enzyme. J Biol Chem *278*, 23180–23186.

Everett, R. D., Meredith, M., Orr, A., Cross, A., Kathoria, M., and Parkinson, J. (1997). A novel ubiquitin-specific protease is dynamically associated with the PML nuclear domain and binds to a herpesvirus regulatory protein. Embo J *16*, 1519–1530.

Febbo, P. G., Richie, J. P., George, D. J., Loda, M., Manola, J., Shankar, S., Barnes, A. S., Tempany, C., Catalona, W., Kantoff, P. W., and Oh, W. K. (2005). Neoadjuvant docetaxel before radical prostatectomy in patients with high-risk localized prostate cancer. Clin Cancer Res *11*, 5233–5240.

Fujiwara, T., Saito, A., Suzuki, M., Shinomiya, H., Suzuki, T., Takahashi, E., Tanigami, A., Ichiyama, A., Chung, C. H., Nakamura, Y., and Tanaka, K. (1998). Identification and chromosomal assignment of USP1, a novel gene encoding a human ubiquitin-specific protease. Genomics *54*, 155–158.

Gousseva, N., and Baker, R. T. (2003). Gene structure, alternate splicing, tissue distribution, cellular localization, and developmental expression pattern of mouse deubiquitinating enzyme isoforms Usp2-45 and Usp2-69. Gene Expr *11*, 163–179.

Graner, E., Tang, D., Rossi, S., Baron, A., Migita, T., Weinstein, L. J., Lechpammer, M., Huesken, D., Zimmermann, J., Signoretti, S., and Loda, M. (2004). The isopeptidase USP2a regulates the stability of fatty acid synthase in prostate cancer. Cancer Cell *5*, 253–261.

Hassiepen U., Eidhoff, U., Meder, G., Bulber, J.F., Hein, A., Bodendorf, U., Lorthiois, E., and Martoglio, B. (2007). A sensitive fluorescence intensity assay for deubiquitinating proteases using ubiquitin-rhodamine110-glycine as substrate. Anal Biochem. *371(2)*, 201–207.

Hellerbrand, C., Bumes, E., Bataille, F., Dietmaier, W., Massoumi, R., and Bosserhoff, A. K. (2007). Reduced expression of CYLD in human colon and hepatocellular carcinomas. Carcinogenesis *28*, 21–27.

Hemelaar, J., Galardy, P. J., Borodovsky, A., Kessler, B. M., Ploegh, H. L., and Ovaa, H. (2004). Chemistry-based functional proteomics: mechanism-based activity-profiling tools for ubiquitin and ubiquitin-like specific proteases. J Proteome Res *3*, 268–276.

Hicke, L. (2001). A new ticket for entry into budding vesicles-ubiquitin. Cell *106*, 527–530.

Hoekstra, D., Tyteca, D., and van, I. S.C. (2004). The subapical compartment: a traffic center in membrane polarity development. J Cell Sci *117*, 2183–2192.

Horn, M., Collingro, A., Schmitz-Esser, S., Beier, C. L., Purkhold, U., Fartmann, B., Brandt, P., Nyakatura, G. J., Droege, M., Frishman, D., et al. (2004). Illuminating the evolutionary history of chlamydiae. Science *304*, 728–730.

Hu, M., Li, P., Li, M., Li, W., Yao, T., Wu, J. W., Gu, W., Cohen, R. E., and Shi, Y. (2002). Crystal structure of a UBP-family deubiquitinating enzyme in isolation and in complex with ubiquitin aldehyde. Cell *111*, 1041–1054.

Hu, M., Li, P., Song, L., Jeffrey, P. D., Chenova, T. A., Wilkinson, K. D., Cohen, R. E., and Shi, Y. (2005). Structure and mechanisms of the proteasome-associated deubiquitinating enzyme USP14. Embo J *24*, 3747–3756.

Hu, M., Gu, L., Li, M., Jeffrey, P. D., Gu, W., and Shi, Y. (2006). Structural basis of competitive recognition of p53 and MDM2 by HAUSP/USP7: implications for the regulation of the p53-MDM2 pathway. PLoS Biol *4*, e27.

Huang, T. T., Nijman, S. M., Mirchandani, K. D., Galardy, P. J., Cohn, M. A., Haas, W., Gygi, S. P., Ploegh, H. L., Bernards, R., and D'Andrea, A. D. (2006). Regulation of monoubiquitinated PCNA by DUB autocleavage. Nat Cell Biol *8*, 339–347.

Iyer, L. M., Burroughs, A. M., and Aravind, L. (2006). The prokaryotic antecedents of the ubiquitin-signaling system and the early evolution of ubiquitin-like beta-grasp domains. Genome Biol *7*, R60.

Jenner, M. W., Leone, P. E., Walker, B. A., Ross, F. M., Johnson, D. C., Gonzalez, D., Chiecchio, L., Dachs Cabanas, E., Dagrada, G. P., Nightingale, M., et al. (2007). Gene mapping and expression analysis of 16q loss of heterozygosity identifies WWOX and CYLD as being important in determining clinical outcome in multiple myeloma. Blood *110*, 3291–3300.

Jin, J., Cardozo, T., Lovering, R. C., Elledge, S. J., Pagano, M., and Harper, J. W. (2004). Systematic analysis and nomenclature of mammalian F-box proteins. Genes Dev *18*, 2573–2580.

Johnson, E. S., and Blobel, G. (1997). Ubc9p is the conjugating enzyme for the ubiquitin-like protein Smt3p. J Biol Chem *272*, 26799–26802.

Johnson, E. S., Schwienhorst, I., Dohmen, R. J., and Blobel, G. (1997). The ubiquitin-like protein Smt3p is activated for conjugation to other proteins by an Aos1p/Uba2p heterodimer. Embo J *16*, 5509–5519.

Johnston, S. C., Larsen, C. N., Cook, W. J., Wilkinson, K. D., and Hill, C. P. (1997). Crystal structure of a deubiquitinating enzyme (human UCH-L3) at 1.8 A resolution. Embo J *16*, 3787–3796.

Johnston, S. C., Riddle, S. M., Cohen, R. E., and Hill, C. P. (1999). Structural basis for the specificity of ubiquitin C-terminal hydrolases. Embo J *18*, 3877–3887.

Kaneko, T., Kumasaka, T., Ganbe, T., Sato, T., Miyazawa, K., Kitamura, N., and Tanaka, N.(2003). Structural insight into modest binding of a non-PXXP ligand to the signal transducing adaptor molecule-2 Src homology 3 domain. J Biol Chem *278*, 48162–48168.

Kato, M., Miyazawa, K., and Kitamura, N. (2000). A deubiquitinating enzyme UBPY interacts with the Src homology 3 domain of Hrs-binding protein via a novel binding motif PX(V/I)(D/N)RXXKP. J Biol Chem *275*, 37481–37487.

Kerscher, O., Felberbaum, R., and Hochstrasser, M. (2006). Modification of proteins by ubiquitin and ubiquitin-like proteins. Annu Rev Cell Dev Biol *22*, 159–180.

Kessler, B. M., Fortunati, E., Melis, M., Pals, C. E., Clevers, H., and Maurice, M. M. (2007). Proteome changes induced by knock-down of the deubiquitylating enzyme HAUSP/USP7. J Proteome Res *6*, 4163–4172.

Kirkin, V., and Dikic, I. (2007). Role of ubiquitin- and Ubl-binding proteins in cell signaling. Curr Opin Cell Biol *19*, 199–205.

Knowles, L. M., and Smith, J. W. (2007). Genome-wide changes accompanying knockdown of fatty acid synthase in breast cancer. BMC Genomics *8*, 168.

Kovalenko, A., Chable-Bessia, C., Cantarella, G., Israel, A., Wallach, D., and Courtois, G. (2003). The tumour suppressor CYLD negatively regulates NF-kappaB signalling by deubiquitination. Nature *424*, 801–805.

Kuhajda, F. P. (2006). Fatty acid synthase and cancer: new application of an old pathway. Cancer Res *66*, 5977–5980.

Kusakabe, T., Maeda, M., Hoshi, N., Sugino, T., Watanabe, K., Fukuda, T., and Suzuki, T. (2000). Fatty acid synthase is expressed mainly in adult hormone-sensitive cells or cells with high lipid metabolism and in proliferating fetal cells. J Histochem Cytochem *48*, 613–622.

Li, M., Chen, D., Shiloh, A., Luo, J., Nikolaev, A. Y., Qin, J., and Gu, W. (2002). Deubiquitination of p53 by HAUSP is an important pathway for p53 stabilization. Nature *416*, 648–653.

Li, M., Brooks, C. L., Kon, N., and Gu, W. (2004). A dynamic role of HAUSP in the p53-Mdm2 pathway. Mol Cell *13*, 879–886.

Li, S. J., and Hochstrasser, M. (2000). The yeast ULP2 (SMT4) gene encodes a novel protease specific for the ubiquitin-like Smt3 protein. Mol Cell Biol *20*, 2367–2377.

Liakopoulos, D., Doenges, G., Matuschewski, K., and Jentsch, S. (1998). A novel protein modification pathway related to the ubiquitin system. Embo J *17*, 2208–2214.

Lin, H., Keriel, A., Morales, C. R., Bedard, N., Zhao, Q., Hingamp, P., Lefrancois, S., Combaret, L., and Wing, S. S. (2000). Divergent N-terminal sequences target an inducible testis deubiquitinating enzyme to distinct subcellular structures. Mol Cell Biol *20*, 6568–6578.

Loda, M., Cukor, B., Tam, S. W., Lavin, P., Fiorentino, M., Draetta, G. F., Jessup, J. M., and Pagano, M. (1997). Increased proteasome-dependent degradation of the cyclin-dependent kinase inhibitor p27 in aggressive colorectal carcinomas. Nat Med *3*, 231–234.

Love, K. R., Catic, A., Schlieker, C., and Ploegh, H. L. (2007). Mechanisms, biology and inhibitors of deubiquitinating enzymes. Nat Chem Biol *3*, 697–705.

Luchansky, S. J., Lansbury, P. T., Jr., and Stein, R. L. (2006). Substrate recognition and catalysis by UCH-L1. Biochemistry *45*, 14717–14725.

Ma, Y. M., Boucrot, E., Villen, J., Affar el, B., Gygi, S. P., Gottlinger, H. G., and Kirchhausen, T. (2007). Targeting of AMSH to endosomes is required for epidermal growth factor receptor degradation. J Biol Chem *282*, 9805–9812.

Makarova, K. S., Aravind, L., and Koonin, E. V. (2000). A novel superfamily of predicted cysteine proteases from eukaryotes, viruses and Chlamydia pneumoniae. Trends Biochem Sci *25*, 50–52.

Massoumi, R., Chmielarska, K., Hennecke, K., Pfeifer, A., and Fassler, R. (2006). Cyld inhibits tumor cell proliferation by blocking Bcl-3-dependent NF-kappaB signaling. Cell *125*, 665–677.

Masuya, D., Huang, C., Liu, D., Nakashima, T., Yokomise, H., Ueno, M., Nakashima, N., and Sumitomo, S. (2006). The HAUSP gene plays an important role in non-small cell lung carcinogenesis through p53-dependent pathways. J Pathol *208*, 724–732.

McCullough, J., Clague, M. J., and Urbe, S. (2004). AMSH is an endosome-associated ubiquitin isopeptidase. J Cell Biol *166*, 487–492.

McCullough, J., Row, P. E., Lorenzo, O., Doherty, M., Beynon, R., Clague, M. J., and Urbe, S. (2006). Activation of the endosome-associated ubiquitin isopeptidase AMSH by STAM, a component of the multivesicular body-sorting machinery. Curr Biol *16*, 160–165.

Melandri, F., Grenier, L., Plamondon, L., Huskey, W. P., and Stein, R. L. (1996). Kinetic studies on the inhibition of isopeptidase T by ubiquitin aldehyde. Biochemistry *35*, 12893–12900.

Minsky, N., and Oren, M. (2004). The RING domain of Mdm2 mediates histone ubiquitylation and transcriptional repression. Mol Cell *16*, 631–639.

Misaghi, S., Galardy, P. J., Meester, W. J., Ovaa, H., Ploegh, H. L., and Gaudet, R. (2005). Structure of the ubiquitin hydrolase UCH-L3 complexed with a suicide substrate. J Biol Chem *280*, 1512–1520.

Miyoshi, Y., Nakayama, S., Torikoshi, Y., Tanaka, S., Ishihara, H., Taguchi, T., Tamaki, Y., and Noguchi, S. (2006). High expression of ubiquitin carboxy-terminal hydrolase-L1 and -L3 mRNA predicts early recurrence in patients with invasive breast cancer. Cancer Sci 97, 523–529.

Mizuno, E., Iura, T., Mukai, A., Yoshimori, T., Kitamura, N., and Komada, M. (2005). Regulation of epidermal growth factor receptor down-regulation by UBPY-mediated deubiquitination at endosomes. Mol Biol Cell 16, 5163–5174.

Mizuno, E., Kobayashi, K., Yamamoto, A., Kitamura, N., and Komada, M. (2006). A deubiquitinating enzyme UBPY regulates the level of protein ubiquitination on endosomes. Traffic 7, 1017–1031.

Mukhopadhyay, D., and Riezman, H. (2007). Proteasome-independent functions of ubiquitin in endocytosis and signaling. Science 315, 201–205.

Naviglio, S., Mattecucci, C., Matoskova, B., Nagase, T., Nomura, N., Di Fiore, P. P., and Draetta, G. F. (1998). UBPY: a growth-regulated human ubiquitin isopeptidase. Embo J 17, 3241–3250.

Neve, R. L. (2003). A new wrestler in the battle between alpha- and beta-secretases for cleavage of APP. Trends Neurosci 26, 461–463.

Niendorf, S., Oksche, A., Kisser, A., Lohler, J., Prinz, M., Schorle, H., Feller, S., Lewitzky, M., Horak, I., and Knobeloch, K. P. (2007). Essential role of ubiquitin-specific protease 8 for receptor tyrosine kinase stability and endocytic trafficking in vivo. Mol Cell Biol 27, 5029–5039.

Nijman, S. M., Luna-Vargas, M. P., Velds, A., Brummelkamp, T. R., Dirac, A. M., Sixma, T. K., and Bernards, R. (2005). A genomic and functional inventory of deubiquitinating enzymes. Cell 123, 773–786.

Oestergaard, V. H., Langevin, F., Kuiken, H. J., Pace, P., Niedzwiedz, W., Simpson, L. J., Ohzeki, M., Takata, M., Sale, J. E., and Patel, K. J. (2007). Deubiquitination of FANCD2 is required for DNA crosslink repair. Mol Cell 28, 798–809.

Oishi, K., Miyazaki, K., Kadota, K., Kikuno, R., Nagase, T., Atsumi, G., Ohkura, N., Azama, T., Mesaki, M., Yukimasa, S., et al. (2003). Genome-wide expression analysis of mouse liver reveals CLOCK-regulated circadian output genes. J Biol Chem 278, 41519–41527.

Oishi, K., Amagai, N., Shirai, H., Kadota, K., Ohkura, N., and Ishida, N. (2005). Genome-wide expression analysis reveals 100 adrenal gland-dependent circadian genes in the mouse liver. DNA Res 12, 191–202.

Osaka, F., Kawasaki, H., Aida, N., Saeki, M., Chiba, T., Kawashima, S., Tanaka, K., and Kato, S. (1998). A new NEDD8-ligating system for cullin-4A. Genes Dev 12, 2263–2268.

Park, K. C., Kim, J. H., Choi, E. J., Min, S. W., Rhee, S., Baek, S. H., Chung, S. S., Bang, O., Park, D., Chiba, T., et al. (2002). Antagonistic regulation of myogenesis by two deubiquitinating enzymes, UBP45 and UBP69. Proc Natl Acad Sci U S A 99, 9733–9738.

Peruzzi, B., and Bottaro, D. P. (2006). Targeting the c-Met signaling pathway in cancer. Clin Cancer Res 12, 3657–3660.

Pickart, C. M., and Eddins, M. J. (2004). Ubiquitin: structures, functions, mechanisms. Biochim Biophys Acta 1695, 55–72.

Pickart, C. M., and Fushman, D. (2004). Polyubiquitin chains: polymeric protein signals. Curr Opin Chem Biol 8, 610–616.

Pickart, C. M., and Rose, I. A. (1985). Ubiquitin carboxyl-terminal hydrolase acts on ubiquitin carboxyl-terminal amides. J Biol Chem 260, 7903–7910.

Polakis, P. (2000). Wnt signaling and cancer. Genes Dev 14, 1837–1851.

Polo, S., Sigismund, S., Faretta, M., Guidi, M., Capua, M. R., Bossi, G., Chen, H., De Camilli, P., and Di Fiore, P. P. (2002). A single motif responsible for ubiquitin recognition and monoubiquitination in endocytic proteins. Nature 416, 451–455.

Popov, N., Herold, S., Llamazares, M., Schulein, C., and Eilers, M. (2007a). Fbw7 and Usp28 regulate myc protein stability in response to DNA damage. Cell Cycle 6, 2327–2331.

Popov, N., Wanzel, M., Madiredjo, M., Zhang, D., Beijersbergen, R., Bernards, R., Moll, R., Elledge, S. J., and Eilers, M. (2007b). The ubiquitin-specific protease USP28 is required for MYC stability. Nat Cell Biol 9, 765–774.

Priolo, C., Tang, D., Brahamandan, M., Benassi, B., Sicinska, E., Ogino, S., Farsetti, A., Porrello, A., Finn, S., Zimmermann, J., et al. (2006). The isopeptidase USP2a protects human prostate cancer from apoptosis. Cancer Res 66, 8625–8632.

Puente, X. S., and Lopez-Otin, C. (2004). A genomic analysis of rat proteases and protease inhibitors. Genome Res 14, 609–622.

Quesada, V., Diaz-Perales, A., Gutierrez-Fernandez, A., Garabaya, C., Cal, S., and Lopez-Otin, C. (2004). Cloning and enzymatic analysis of 22 novel human ubiquitin-specific proteases. Biochem Biophys Res Commun 314, 54–62.

Reiley, W., Zhang, M., and Sun, S. C. (2004). Negative regulation of JNK signaling by the tumor suppressor CYLD. J Biol Chem 279, 55161–55167.

Reiley, W. W., Zhang, M., Jin, W., Losiewicz, M., Donohue, K. B., Norbury, C. C., and Sun, S. C. (2006). Regulation of T cell development by the deubiquitinating enzyme CYLD. Nat Immunol 7, 411–417.

Renatus, M., Parrado, S. G., D'Arcy, A., Eidhoff, U., Gerhartz, B., Hassiepen, U., Pierrat, B., Riedl, R., Vinzenz, D., Worpenberg, S., and Kroemer, M. (2006). Structural basis of ubiquitin recognition by the deubiquitinating protease USP2. Structure 14, 1293–1302.

Rolen, U., Kobzeva, V., Gasparjan, N., Ovaa, H., Winberg, G., Kisseljov, F., and Masucci, M. G. (2006). Activity profiling of deubiquitinating enzymes in cervical carcinoma biopsies and cell lines. Mol Carcinog 45, 260–269.

Rolfe, M., Chiu, M. I., and Pagano, M. (1997). The ubiquitin-mediated proteolytic pathway as a therapeutic area. J Mol Med 75, 5–17.

Ross, C. A., and Pickart, C. M. (2004). The ubiquitin-proteasome pathway in Parkinson's disease and other neurodegenerative diseases. Trends Cell Biol 14, 703–711.

Rossi, S., Graner, E., Febbo, P., Weinstein, L., Bhattacharya, N., Onody, T., Bubley, G., Balk, S., and Loda, M. (2003). Fatty acid synthase expression defines distinct molecular signatures in prostate cancer. Mol Cancer Res 1, 707–715.

Row, P. E., Prior, I. A., McCullough, J., Clague, M. J., and Urbe, S. (2006). The ubiquitin isopeptidase UBPY regulates endosomal ubiquitin dynamics and is essential for receptor down-regulation. J Biol Chem 281, 12618–12624.

Santamaria, D., and Ortega, S. (2006). Cyclins and CDKS in development and cancer: lessons from genetically modified mice. Front Biosci 11, 1164–1188.

Scheel, H., Tomiuk, S., and Hofmann, K. (2003). Elucidation of ataxin-3 and ataxin-7 function by integrative bioinformatics. Hum Mol Genet 12, 2845–2852.

Schwienhorst, I., Johnson, E. S., and Dohmen, R. J. (2000). SUMO conjugation and deconjugation. Mol Gen Genet 263, 771–786.

Sebastiani, V., Visca, P., Botti, C., Santeusanio, G., Galati, G. M., Piccini, V., Capezzone de Joannon, B., Di Tondo, U., and Alo, P. L. (2004). Fatty acid synthase is a marker of increased risk of recurrence in endometrial carcinoma. Gynecol Oncol 92, 101–105.

Setsuie, R., and Wada, K. (2007). The functions of UCH-L1 and its relation to neurodegenerative diseases. Neurochem Int 51, 105–111.

Sheng, Y., Saridakis, V., Sarkari, F., Duan, S., Wu, T., Arrowsmith, C. H., and Frappier, L. (2006). Molecular recognition of p53 and MDM2 by USP7/HAUSP. Nat Struct Mol Biol 13, 285–291.

Sherr, C. J. (1996). Cancer cell cycles. Science 274, 1672–1677.

Sobhian, B., Shao, G., Lilli, D. R., Culhane, A. C., Moreau, L. A., Xia, B., Livingston, D. M., and Greenberg, R. A. (2007). RAP80 targets BRCA1 to specific ubiquitin structures at DNA damage sites. Science 316, 1198–1202.

Steffan, J. S., Agrawal, N., Pallos, J., Rockabrand, E., Trotman, L. C., Slepko, N., Illes, K., Lukacsovich, T., Zhu, Y. Z., Cattaneo, E., et al. (2004). SUMO modification of Huntingtin and Huntington's disease pathology. Science 304, 100–104.

Stegmeier, F., Rape, M., Draviam, V. M., Nalepa, G., Sowa, M. E., Ang, X. L., McDonald, E. R., 3rd, Li, M. Z., Hannon, G. J., Sorger, P. K., et al. (2007). Anaphase initiation is regulated by antagonistic ubiquitination and deubiquitination activities. Nature 446, 876–881.

Stein, R. L., Chen, Z., and Melandri, F. (1995). Kinetic studies of isopeptidase T: modulation of peptidase activity by ubiquitin. Biochemistry 34, 12616–12623.

Stevenson, L. F., Sparks, A., Allende-Vega, N., Xirodimas, D. P., Lane, D. P., and Saville, M. K. (2007). The deubiquitinating enzyme USP2a regulates the p53 pathway by targeting Mdm2. Embo J 26, 976–986.

Storch, K. F., Lipan, O., Leykin, I., Viswanathan, N., Davis, F. C., Wong, W. H., and Weitz, C. J. (2002). Extensive and divergent circadian gene expression in liver and heart. Nature 417, 78–83.

Su, Y., Ishikawa, S., Kojima, M., and Liu, B. (2003). Eradication of pathogenic beta-catenin by Skp1/Cullin/F box ubiquitination machinery. Proc Natl Acad Sci U S A 100, 12729–12734.

Sun, Y. (2003). Targeting E3 ubiquitin ligases for cancer therapy. Cancer Biol Ther 2, 623–629.

Swinnen, J. V., Brusselmans, K., and Verhoeven, G. (2006). Increased lipogenesis in cancer cells: new players, novel targets. Curr Opin Clin Nutr Metab Care 9, 358–365.

Szymkiewicz, I., Shupliakov, O., and Dikic, I. (2004). Cargo- and compartment-selective endocytic scaffold proteins. Biochem J 383, 1–11.

Taylor, F. R., Bixler, S. A., Budman, J. I., Wen, D., Karpusas, M., Ryan, S. T., Jaworski, G. J., Safari-Fard, A., Pollard, S., and Whitty, A. (1999). Induced fit activation mechanism of the exceptionally specific serine protease, complement factor D. Biochemistry 38, 2849–2859.

Tirat, A., Schilb, A., Riou, V., Leder, L., Gerhartz, B., Zimmermann, J., Worpenberg, S., Eidhoff, U., Freuler, F., Stettler, T., et al. (2005). Synthesis and characterization of fluorescent ubiquitin derivatives as highly sensitive substrates for the deubiquitinating enzymes UCH-L3 and USP-2. Anal Biochem 343, 244–255.

Tran, H. J., Allen, M. D., Lowe, J., and Bycroft, M. (2003). Structure of the Jab1/MPN domain and its implications for proteasome function. Biochemistry 42, 11460–11465.

Trompouki, E., Hatzivassiliou, E., Tsichritzis, T., Farmer, H., Ashworth, A., and Mosialos, G. (2003). CYLD is a deubiquitinating enzyme that negatively regulates NF-kappaB activation by TNFR family members. Nature 424, 793–796.

Valero, R., Bayes, M., Francisca Sanchez-Font, M., Gonzalez-Angulo, O., Gonzalez-Duarte, R., and Marfany, G. (2001). Characterization of alternatively spliced products and tissue-specific isoforms of USP28 and USP25. Genome Biol 2, RESEARCH0043.

van Delft, S., Govers, R., Strous, G. J., Verkleij, A. J., and van Bergen en Henegouwen, P. M. (1997). Epidermal growth factor induces ubiquitination of Eps15. J Biol Chem 272, 14013–14016.

van der Horst, A., de Vries-Smits, A. M., Brenkman, A. B., van Triest, M. H., van den Broek, N., Colland, F., Maurice, M. M., and Burgering, B. M. (2006). FOXO4 transcriptional activity is regulated by monoubiquitination and USP7/HAUSP. Nat Cell Biol 8, 1064–1073.

van der Knaap, J. A., Kumar, B. R., Moshkin, Y. M., Langenberg, K., Krijgsveld, J., Heck, A. J., Karch, F., and Verrijzer, C. P. (2005). GMP synthetase stimulates histone H2B deubiquitylation by the epigenetic silencer USP7. Mol Cell 17, 695–707.

Vasiljeva, O., Reinheckel, T., Peters, C., Turk, D., Turk, V., and Turk, B. (2007). Emerging roles of cysteine cathepsins in disease and their potential as drug targets. Curr Pharm Des 13, 387–403.

Visca, P., Sebastiani, V., Botti, C., Diodoro, M. G., Lasagni, R. P., Romagnoli, F., Brenna, A., De Joannon, B. C., Donnorso, R. P., Lombardi, G., and Alo, P. L. (2004). Fatty acid synthase (FAS) is a marker of increased risk of recurrence in lung carcinoma. Anticancer Res 24, 4169–4173.

Wertz, I. E., O'Rourke, K. M., Zhou, H., Eby, M., Aravind, L., Seshagiri, S., Wu, P., Wiesmann, C., Baker, R., Boone, D. L., et al. (2004). De-ubiquitination and ubiquitin ligase domains of A20 downregulate NF-kappaB signalling. Nature 430, 694–699.

Wilkinson, K. D. (1997). Regulation of ubiquitin-dependent processes by deubiquitinating enzymes. Faseb J 11, 1245–1256.

Wing, S. S. (2003). Deubiquitinating enzymes-the importance of driving in reverse along the ubiquitin-proteasome pathway. Int J Biochem Cell Biol 35, 590–605.

Wright, A., Reiley, W. W., Chang, M., Jin, W., Lee, A. J., Zhang, M., and Sun, S. C. (2007). Regulation of early wave of germ cell apoptosis and spermatogenesis by deubiquitinating enzyme CYLD. Dev Cell 13, 705–716.

Wu, X., Yen, L., Irwin, L., Sweeney, C., and Carraway, K. L., 3rd (2004). Stabilization of the E3 ubiquitin ligase Nrdp1 by the deubiquitinating enzyme USP8. Mol Cell Biol 24, 7748–7757.

Xue, L., Igaki, T., Kuranaga, E., Kanda, H., Miura, M., and Xu, T. (2007). Tumor suppressor CYLD regulates JNK-induced cell death in Drosophila. Dev Cell 13, 446–454.

Yang, J. M. (2007). Emerging roles of deubiquitinating enzymes in human cancer. Acta Pharmacol Sin 28, 1325–1330.

Young, A. L., Kellermayer, R., Szigeti, R., Teszas, A., Azmi, S., and Celebi, J. T. (2006). CYLD mutations underlie Brooke-Spiegler, familial cylindromatosis, and multiple familial trichoepithelioma syndromes. Clin Genet 70, 246–249.

Zhang, D., Zaugg, K., Mak, T. W., and Elledge, S. J. (2006a). A role for the deubiquitinating enzyme USP28 in control of the DNA-damage response. Cell 126, 529–542.

Zhao, W., Kridel, S., Thorburn, A., Kooshki, M., Little, J., Hebbar, S., and Robbins, M. (2006). Fatty acid synthase: a novel target for antiglioma therapy. Br J Cancer 95, 869–878.

Zhong, S., Salomoni, P., and Pandolfi, P. P. (2000). The transcriptional role of PML and the nuclear body. Nat Cell Biol 2, E85–E90.

Proteolysis Targeting Chimeric Molecules: Recruiting Cancer-Causing Proteins for Ubiquitination and Degradation

Agustin Rodriguez-Gonzalez and Kathleen M. Sakamoto

Abstract Protein degradation is one of the mechanisms employed by the cell for irreversibly destroying proteins. In eukaryotes, ATP-dependent protein degradation in the cytoplasm and nucleus is carried out by the 26S proteasome. Most proteins are targeted to the 26S proteasome by covalent attachment of a multiubiquitin chain. A key component of the enzyme cascade that results in attachment of the multiubiquitin chain to the target or labile protein is the E3 ubiquitin ligase that controls the specificity of the ubiquitination reaction. Defects in ubiquitin-dependent proteolysis have been shown to result in a variety of human diseases, including cancer, neurodegenerative diseases, and metabolic disorders. We have developed a novel approach to target proteins that cause cancer for ubiquitination and degradation. This technology, known as Protac, involves a chimeric molecule that could potentially recruit any cancer-causing protein to an E3 ligase for ubiquitination and subsequent degradation. In this chapter, we describe the development of this technology for cancer therapy.

Keywords Protein degradation • Ubiquitination • Cancer therapy • Breast cancer • Prostate cancer • Chimeric molecule • Protac

Introduction

Many biological pathways in cancer are aberrant because of the overexpression or hyperactivity of a protein. Approximately 500 genes and their encoded proteins have been related directly with the transformation process (Huret 2008). Proteins such as transcription factors, steroid receptors, small GTPases, phosphatases, kinases, cyclins, and ubiquitin ligases have altered expression and function during cellular

A.Rodriguez-Gonzalez and K.M. Sakamoto (✉)
Department of Pediatrics, Department of Pathology & Laboratory Medicine,
Gwynne Hazen Cherry Laboratories, Jonsson Comprehensive Cancer Center,
David Geffen School of Medicine at UCLA, Los Angeles, CA, 90095, USA

K. Sakamoto and E. Rubin (eds.), *Modulation of Protein Stability in Cancer Therapy*,   147
DOI: 10.1007/978-0-387-69147-3_9, © Springer Science+Business Media, LLC 2009

transformation. Since the initial identification and categorization of cancer, numerous approaches to study and correct these abnormal pathways have been developed. Initially, the paradigm for drug discovery was *"from molecule to target"* , which resulted in the creation of natural compound libraries, followed by the production of synthetic sources (Gaither 2007). The availability of thousands of compounds allowed for screens to identify relationships between molecular families and desired phenotypes. However, screening of these compounds faces many obstacles. The lack of knowledge in the complexity of targets, and the possibility that a potential molecule may bind to multiple targets, has made the compound screening method a way of obtaining therapeutic drugs that present side effects and a not totally understood mechanism of action (Gaither 2007). Even in the most directed screening against one single protein, specific molecules that are recovered must pass the scrutiny of many rounds of optimizations to increase the compound activity, permeability, and clearance resistance. Unfortunately, despite these approaches, many compounds still bind to multiple proteins and therefore elicit nonspecific "off-target" effects (Copeland et al. 2006; Gaither 2007; Stumpfe et al. 2007). Lately, chemoinformatic approaches have allowed new approaches to screen compounds. Interactions between targeted proteins and compounds can be predicted in virtual screens by silico alignment and modeling algorithms (Kitchen et al. 2004). Still, these approaches do not preclude the possibility of binding to multiple targets.

In the late 1990s, there was a switch *"from target to molecule"* rather than *"from molecule to target."* This change in the field of drug discovery resulted from the development of RNA interference technology (RNAi). RNAi induces the selective degradation of mRNA that codes for target proteins, inducing a knock down in a single gene at the RNA level (Fire et al. 1998). This approach allowed scientists to relate various phenotypes with the inhibition of a single protein, thus enabling the design of therapeutic strategies based on RNAi technology. The specificity of this new approach was based on the sequence of the targeted mRNA offering high specificity in the down regulation of the protein of interest. This approach improved target identification because the effect caused by protein down-regulation is target-centric rather than molecule-centric (Gaither 2007). The mechanistic implications of RNAi in a therapeutic approach can be extended in vivo using shRNA vectors. Live cells can be transduced with retroviral or lentiviral shRNAs, thereby integrating the shRNA in the genome with greater efficacy (Borawski et al. 2007; Root et al. 2006), and inducing a knockdown of a specific target in animal models (Zhu et al. 2007).

While small molecule and RNAi therapies have provided a large number of potentially beneficial compounds, both approaches have significant drawbacks. For most small molecules, off-target effects have been described at high concentrations, e.g., high micromolar range, resulting in toxicity and decreased tolerance by patients. When selecting the sequences for RNAi technology, careful attention must be focused to reduce cross-reactivity to homologous genes (Birmingham et al. 2006; Gaither 2007). In cancer, the RNAi approach has to overcome the possibilities of random mismatch in the mRNA because of DNA instability. The possibility of a single mismatch in the targeted sequence might eliminate the potency of the shRNA (Gaither 2007). Another concern in the RNAi approach is the induction of an interferon response and stress signaling by the exogenous double-stranded siRNA (Heidel et al. 2004), generating a different phenotype from what is expected from knocking down the desired protein.

While both compound screens and RNAi present off-target effects, the first approach provides a temporal control, while the second approach is more stable and long-acting. To resolve the specificity concern, it was proposed to use RNAi paradigm at the protein level (Sakamoto et al. 2001). This new approach would maintain the *"from target to molecule"* paradigm, similar to RNAi technology. Yet, it maintains the advantage of affecting cells at the protein level, thus avoiding the mismatch problem. This approach exploits the ubiquitin-proteasome system (UPS) as the molecular mechanism of protein inhibition (Sakamoto et al. 2001).

Ubiquitin-Proteasome System

The steady-state level of proteins in the cell depends on their synthesis and degradation. The majority of protein degradation occurs through the UPS. These proteins are tagged for degradation by multimers of ubiquitin that the 26S proteasome recognizes and removes (Nandi et al. 2006). Ubiquitin is a conserved protein, composed of 76 amino acids, covalently bound to the tagged protein via the ε-amino group of a lysine by an isopeptide union to the carboxy terminal glycine of the ubiquitin. More isopeptide linkages are formed between the carboxy terminus of the ubiquitin with an ε-amino group of the following ubiquitin, forming a chain of ubiquitins (Ciechanover and Iwai 2004; Pickart 2001; Weissman 2001). There are different steps and elements that contribute to this pathway. These include ubiquitin ligases (E3), ubiquitin carriers or conjugating enzymes (E2), and ubiquitin activator enzymes (E1) (Fig. 1). Ubiquitin is activated by an E1 enzyme which generates ubiquitin-E1 thiol ester as an activated form of ubiquitin in an ATP-dependent reaction. This ubiquitin-E1 intermediate can be recognized by different E2 enzymes and the ubiquitin is transferred again by another thiol ester linkage. It has been estimated that there are few E1s, dozens of E2 and hundreds of E3 ligases (Semple 2003). The timing and selection of which proteins are labeled with ubiquitin is established by the combination of the E2 and E3 enzyme activity. These enzymatic reactions are responsible for the fine-tuning and specificity of the UPS that denotes which proteins must be ubiquitinated (Ciechanover and Iwai 2004; Pickart 2001; Pickart and Cohen 2004; Voges et al. 1999; Weissman 2001).

There are several ways which trigger the recognition of a protein for ubiquitination by the E3 ligase. First, recognition depends on the presence of special sequences in the proteins (e.g., destruction box in cyclins), the amine terminal residue (also known as N-end rule – basic amino acids make the proteins less stable), the display of a hydrophobic sequence, or post-translation modifications (e.g., phosphorylation, hydroxylation, or oxidation)(Varshavsky 2005). Once the protein has been poly-ubiquitinated, the 26S proteasome recognizes the ubiquitin-tagged protein and starts the unfolding process followed by the proteolysis. The proteins are initially

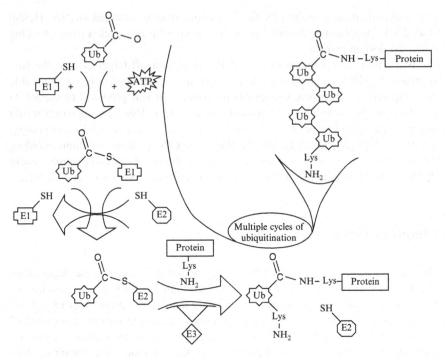

Fig. 1 Ubiquitin-proteasome system: Ubiquitin is activated by an E1 enzyme carrying ubiquitin-E1 thiol ester as an activated form of ubiquitin in an ATP-dependent reaction. This ubiquitin-E1 intermediate can be recognized by different E2 enzymes and the ubiquitin is transferred again by another thiol ester linkage. After multiple cycles of ubiquitination, proteasome recognizes the polyubiquitin tail and degrades the protein

unfolded, deubiquitinated, and translocated into the proteolytic chambers in an ATP-dependent way. Then the polypeptide is degraded in an ATP-independent way by carboxypeptidases, aminopeptidases, and endopeptidases in the proteolytic chamber (Chandu and Nandi 2004). The mammalian proteasomes are localized in the cytosol and nucleus. The 26S proteasome is formed by the 20S proteasome (cylindrical proteolytic chamber) capped on both sides by the 19S. The 20S unit harbors ATP-independent peptidase activity (chymotrypsin-like activity, trypsin-like activity and caspase-like activity). The 19S units unfolds, removes the polyubiquitin tag, and channels the proteins to the proteolytic chamber in an ATP-dependent process (Glickman et al. 1998; Pickart and Cohen 2004; Voges et al. 1999).

This pathway of protein degradation is present and active in every cell at every phase of the cell cycle. In theory, it is capable of inducing a "knock down" of any protein at any given time. The idea is to design a protein degradation inducer that would act as a specific bridge between the targeted protein and the UPS by recruiting the activity of the ubiquitous UPS elements against a targeted protein. This new technology is called Protac (proteolysis tArgeting chimeric molecule) (Sakamoto et al. 2001).

The Protac Concept (Moving to the Protein Level Paradigm *"Target to Molecule"*)

Protacs are "bridging molecules" that can link together a disease-related protein to an E3 ligase (Sakamoto 2005). The Protac consists of two minimum moieties; the first is a peptidic sequence (destruction box) that is recognized by an E3 ubiquitin ligase, and the second moiety is the natural ligand of the desired protein target (Fig. 2a). Both the moieties are chemically and covalently linked to constitute a chimeric heterobifunctional small molecule. On one end, it binds specifically to the desired protein target and on the other end, it recruits the E3 ubiquitin ligase (Fig. 2b).

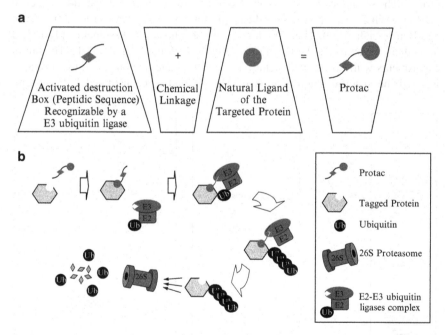

Fig. 2 Protac concept. (**a**) Molecule scheme. (●) Ligand of the targeted protein. (✦) Peptidic sequence recognizable by an E3 ubiquitin ligase. (✦●)Protac "bridging molecule" is the result of linking chemically the ligand and the peptidic sequence. (**b**) This Protac is a "bringing molecule" that can link together cancer-related protein and an E3 ligase. The ligand part of the molecule would interact with the targeted protein, while the peptidic sequence of the Protac would recruit the E3 ligase complex. This E3 ligase complex would link ubiquitin molecules to the targeted protein

Protac (Principle of Proof)

As a starting point in the design, construction, and assay of the first Protac, a target with a well-characterized small molecule ligand was tested. Methionine aminopeptidase-2 (MetAP-2) is a protein that cleaves the N-terminal methionine from newly formed polypeptides. It is the therapeutic target of angiogenesis inhibitors such as fumagillin and ovalicin (Griffith et al. 1997; Sin et al. 1997). Ovalicin inhibits MetAP-2 by *covalently* binding at His-231 of its active site. This inhibition leads to subsequent inhibition of endothelial cell proliferation (Yeh et al. 2000). This was a model that could be used to test Protacs under optimal conditions. Hence, with MetAP-2 as the target, ovalicin could be the well-characterized ligand of the target protein. The next step was to identify a peptidic sequence that would be recognized by an E3 ligase. MetAP-2 is a protein that is neither ubiquitinated nor an endogenous substrate of the E3 ligase complex, called SCF$^{\beta\text{-TCR}}$ (Skp1-Cullin-F-box-Hrt1), which made this a convenient E3 ubiquitin ligase to be recruited by Protac. On the other hand, it was previously shown that SCF$^{\beta\text{-TCR}}$ binds and ubiquitinates IKBα protein. The substrate recognition was based on a minimal phosphopeptide sequence "DRHDS*GLDS*M" (asterisks indicate phosphoserines) (Ben-Neriah 2002; Karin and Ben-Neriah 2000). This 10-amino acid phosphopetide was chemically linked to ovalicin, constituting the first Protac (Fig. 3) (Sakamoto et al. 2001). The initial experiments with Protac were achieved "in vitro," and demonstrated that Protac 1 induced the specific ubiquitination of MetAP-2 in a reconstituted UPS system.

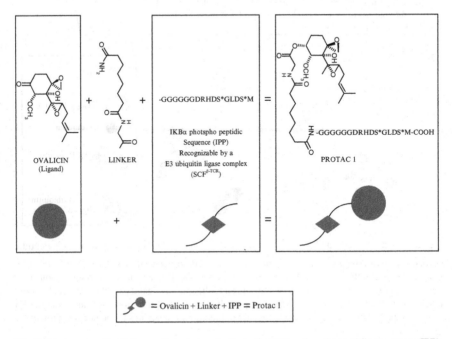

Fig. 3 First protac. Ovalicin was linked chemically to the IKBα phosphopeptidic sequence (IPP)

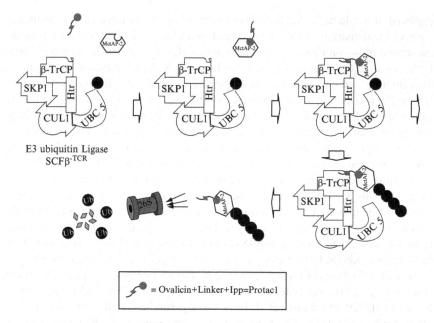

Fig. 4 Ovalicin interacts with MetAP-2, while the phosphopeptidic sequence of Protac (IPP) is recognized by the E3 ligase complex SCF$^{β-TCR}$. Then, SCF$^{β-TCR}$ links ubiquitin molecules to MetAP-2 protein and finally the proteasome recognizes the ubiquitin tail and degrades the protein

Furthermore, this ubiquitination was recognized by the endogenous 26S proteasomes in *Xenopus* egg extracts (Sakamoto et al. 2001). Results demonstrated that ovalicin binds to MetAP-2 and the IKBα phosphopeptide sequence (IPP) binds to SCF$^{β-TCR}$. These interactions occur in close enough proximity such that MetAP-2 is ubiquitinatinated (Fig. 4). This was the first proof of concept demonstrating that Protac technology could be used to chemically knock down the expression of desired targets at the protein level (Sakamoto et al. 2001).

Making New and Improved Protacs

After demonstrating the feasibility of Protac 1 through *covalent* interactions between the ligand portion and the desired target, the next milestone was to design a Protac that could target proteins through noncovalent interactions. The design of a second generation Protac would increase the versatility and utility of Protacs to induce degradation of proteins for cancer, since one molecule of Protac would have the ability to induce the ubiquitination of more than one protein, following a "*hit and run*" pattern. A Protac with a noncovalent interaction with the desired target would increase the efficacy of degradation requiring less Protac. Since Protac has been considered as a possible approach to cancer therapy, new protein targets need to be explored. It would be desirable for these new proteins to be targeted in high-impact variety of malignancies such as prostate and breast tumors. Prostate cancer represents the second most common type of cancer in men (Culig et al. 2003; Montironi 2001; Parkin et al. 2005). One in six men will be diagnosed with prostate cancer during their lifetime, and one in thirty will die because of metastatic disease (Nelson et al. 2003). The association of the androgen receptor (AR) with pathogenesis of prostate cancer has made it a suitable target for Protac development. Once the disease has been diagnosed, there are four treatments: radical prostatectomy, external beam radiation, brachytherapy, cryotherapy, or hormonal therapy. The first four methods are very effective in local disease, but only hormone therapy is effective for systemic disease. At this time, androgen deprivation is the only therapy for treating metastatic prostate cancer. Initially 85% of metastatic prostate tumors respond to hormone therapy, which consists of inhibition of AR activation by the reduction of the androgen levels in the serum or direct receptor inhibition by a chemical antagonist. Eventually, hormone refractory disease develops within 18–24 months and most of these patients will die in 1–2 years (Santen 1992; Savarese et al. 2001). In 50% of the hormone-refractory cases, molecular bases of resistance are due to changes in AR activation or expression (Bakin et al. 2003; Bubendorf et al. 1999; Craft et al. 1999; Culig et al. 1994; Franco et al. 2003; Gaddipati et al. 1994; Godoy-Tundidor et al. 2002; Taplin et al. 1999; Ueda et al. 2002). These changes are consequences of AR overexpression, mutation, or direct activation by crosstalking with other pathways. In all of these cases, the presence and activation of AR is required (Eder et al. 2002; Liao et al. 2005; Linja et al. 2001; Zegarra-Moro et al. 2002). Protac was a logical approach to treat hormone refractory cases, since Protac would induce the specific degradation of the receptor, which would prevent its activation and consequently its downstream signaling. Another similar experimental model would be breast cancer. With 1.15 million new cases per year and 411,000 annual deaths, breast cancer is the most common cause of cancer mortality in women (Parkin et al. 2005). There are intriguing similarities with prostate cancer, in that 70% of breast tumors express estradiol receptor α (ERα) and respond initially to hormone therapy. Resistance develops from estrogenicity of the drugs (Hoffmann and Sommer 2005) or cross-activation with epidermal growth factor pathway (Shou et al. 2004). Equal to the prostate model, Protac is an appealing approach in hormone refractory cases of breast cancer, since Protac would induce the specific degradation of ERα, which would prevent its activation and consequently its mitotic signaling. To this end, two other Protacs were designed (Sakamoto et al. 2003); one to target the estradiol receptor (ERα) and the other to target the androgen receptor (AR), namely, Protacs 2 and 3, respectively. The ligands were estradiol (E2) and dihydrotestosterone (DHT) for Protacs 2 and 3, correspondingly. The peptidic sequence used for Protacs 2 and 3 was the same as that used in Protac 1; the IkBa phosphopeptide, IPP, that binds to SCF$^{\beta\text{-TCR}}$ (Fig. 5, 6). E2 and DHT are the physiologic ligands for ER and AR and bind through noncovalent interactions. In the case of Protac 2, specific induction of ER ubiquitination was observed in vitro with a UPS-reconstituted system, as was seen in the case of Protac 1, displaying its activity at concentrations ranging from 0.1–10 µM (Sakamoto et al. 2003). These ER-ubiquitinated intermediates were also recognized by purified yeast

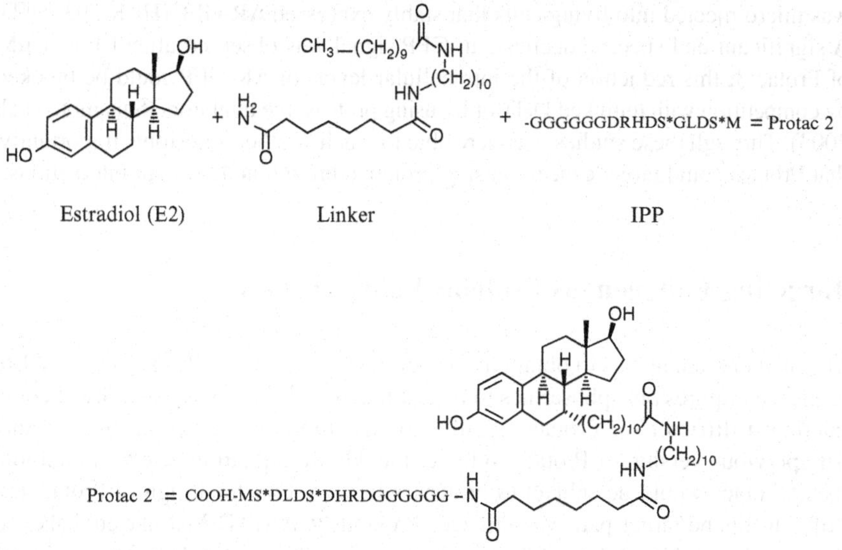

Fig. 5 Protac 2. Estradiol (E2) was linked chemically to the IKBα phosphopeptidic sequence (IPP). So, E2 would bind to estradiol receptor and IPP would recruit the SCF$^{β-TCR}$ ubiquitin ligase activity

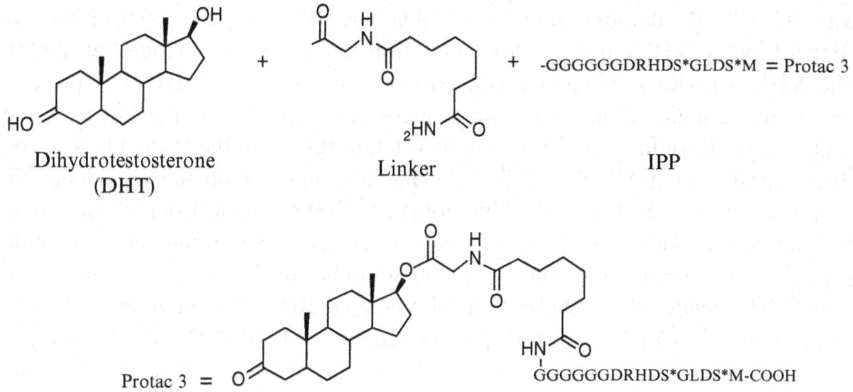

Fig. 6 Protac 3. Dihydrotestosterone (DHT) was linked chemically to the IKBα phosphopeptidic sequence (IPP). So, DHT would bind to androgen receptor and IPP would recruit the SCF$^{β-TCR}$ ubiquitin ligase activity

26S proteasomes, resulting in complete degradation of the ubiquitinated ER proteins (Sakamoto et al. 2003). These experiments provided the evidence that Protacs could recruit and induce the ubiquitination of targeted proteins through noncovalent interactions, allowing for the possibility of one molecule of Protac to initiate the degradation of multiple molecules of proteins involved in the pathogenesis of cancer. Protac 3 (DHT-based) was then tested in a more dynamic context. It was designed to target the AR; therefore, using DHT as the ligand was a logical choice (Fig. 6). Protac 3 was microinjected into living cells that stably expressed AR-GFP (HEK 293[AR-GFP]). A significant and specific decrease in GFP signal was observed after 1 h at 1 μM of Protac 3; this reduction of the intracellular levels of AR-GFP could be blocked by competition with unlinked DHT or by using proteasome inhibitors (Sakamoto et al. 2003). Through these studies, we were able to reach another milestone, importantly, that Protacs could target cancer-causing proteins through noncovalent interactions.

Targeting Endogenous Proteins Using Protacs

The next challenge was to obtain a cell-permeable version of Protac since the IPP sequence requires phosphoserines to recruit the SCF$^{\beta-TCR}$ E3 ligase activity, thereby making it difficult for Protacs 2 and 3 to pass through the cellular membrane. The previous version of Protacs to target the AR were microinjected. In addition, these Protacs would be subject to proteases in vivo. For the design of Protac 4, a different ligand/target pair was utilized. Previously, ARIAD Pharmaceuticals had developed a F36V mutation of the FK506-binding protein (FKBP12). This molecule generates a "hole" into which the AP21998 fits via a hydrophobic "bump," which allowed for specificity of this particular ligand to the mutant FKBP over the wild-type protein (Rollins et al. 2000; Yang et al. 2000). The peptide sequence in this case was "ALAPYIP;" the minimum domain that was shown to be recognized by von Hippel-Lindau (VHL), part of an E3 ubiquitin ligase complex (Hon et al. 2002). The VHL peptide is within the hypoxia-inducible factor-1α (HIF-1α) protein. In normoxic conditions, this transcription factor is hydroxylated at P564 by a proline hydroxylase (Epstein et al. 2001). Hydroxylation results in the recognition of the ALAPYIP domain in HIF 1α by VHL, leading to ubiquitination and degradation by the proteasome (Ohh et al. 2000; Tanimoto et al. 2000). The last step of the Protac 4 design was to add a poly-D-arginine tag on the carboxy terminus of the peptide sequence, to increase the permeability and solubility of the molecule (Kirschberg et al. 2003; Wender et al. 2000). Protac 4 was synthesized by combining AP21998 (specific ligand of FKBP12) to the peptide sequence ALAPYIP (that is recognized by VHL E3 ligase) and the addition of a polyarginine tail that would increase its permeability (Fig. 7) (Schneekloth et al. 2004). The activity of this new Protac 4 was assayed in HeLa cells that expressed the mutant FKBP12 fused to EGFP. It was observed that the protein levels of FKBP12 were reduced after 2.5 h of cell treatment with 25 μM Protac 4 (Schneekloth et al. 2004). The cell viability did not appear to be affected following Protac treatment. Finally, Protac 5, with the ALAPYIP peptide sequence but a different ligand (DHT to target AR), recruited VHL and allows for increased permeability (Fig. 8). This Protac was tested in HEK 293 cells that stably expressed AR-GFP (kindly provided by Charles Sawyers). Protac 5 was bath-applied in culture to HEK293 and not microinjected as with the prevous version. A significant and specific decrease in AR-GFP expression was observed after 1 h of treatment at 25 μM (Schneekloth et al. 2004). These results indicate that Protac technology could be used in a dynamic context and without permeability

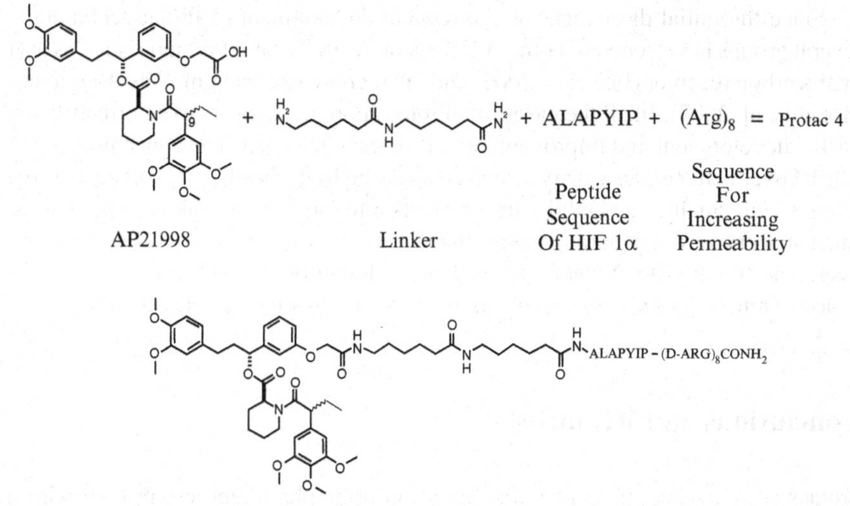

AP21998   Linker   Peptide Sequence Of HIF 1α   Sequence For Increasing Permeability

Protac 4

Fig. 7 Protac 4. AP21998 was linked chemically to the peptidic sequence of HIF1α and finally a poly-D-arginine tag was added in the peptidic side of the Protac. The poly-D-arginine peptide would allow Protac to pass the membrane of the cell, AP21998 would bind to FKBP12 and finally the HIF1α peptidic sequence would recruit the VHL ubiquitin ligase activity

DHT   Linker   Peptide Sequence Of HIF 1α   Sequence For Increasing Permeability

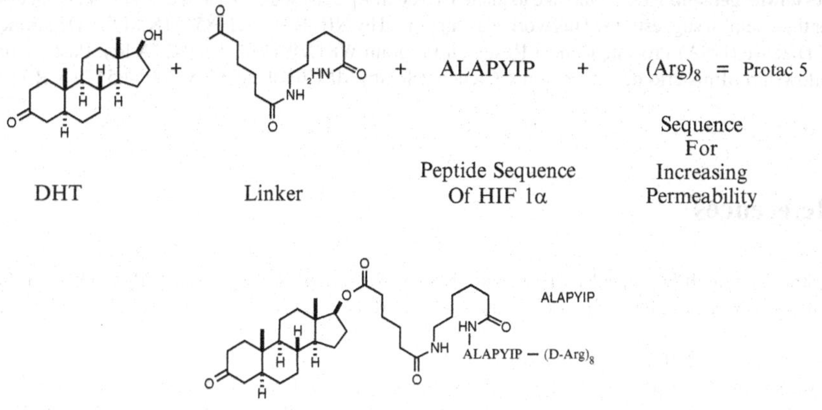

Protac 5

Fig. 8 Protac 5. DHT was linked chemically to the peptidic sequence of HIF1α and finally a poly-D-arginine tag was added in the peptidic side of the Protac. The poly-D-arginine peptide would allow Protac to pass the membrane of the cell, DHT would bind to androgen receptor, and finally the HIF1α peptidic sequence would recruit the VHL ubiquitin ligase activity

issues. The versatility of Protac technology allows for the use of ligands derived from nature (e.g., DHT and E2) or syntheses (e.g., ovalicin and AP21998) to degrade cancer-causing proteins.

Since the initial discovery and subsequent development of Protac technology, several groups have generated other versions of Protac to target proteins such as aryl hydrocarbon receptor (Lee et al. 2007) and in other breast cancer models (Bargagna-Mohan et al. 2005). In all the cases, the Protac approach appears to be effective but further development and improvement in Protacs are needed. The concentrations at which Protac demonstrate activity in living cells are high and therefore it is very expensive to transform small-scale cellular experiments into large-scale animal experiments. Current work in our laboratory and others are toward testing new approaches to overcome this hurdle. Newer approaches are being developed to identify small-molecule binding partners that recognize either the E3 ligase or the cancer target.

Conclusions and Remarks

Protacs are chimeric molecules that could, in principle, target any protein with a known ligand or binding partner. Given the lack of specificity of currently used small-molecule inhibitors for cancer therapy, Protacs provide a new approach to target specific cancer-promoting proteins for ubiquitination and degradation. Further development of Protacs are needed to improve permeability, potency, and large-scale synthesis of these versatile compounds.

Acknowledgements We would like to thank Dr. Raymond J. Deshaies, Alan Ikeda and Dejah Judelson for their helpful suggestions. This work was supported by NIH R21 CA108545 (K.M.S.), Department of Defense (USA) Prostate Cancer Research Program W81XWH-06-1-0192 (A.R.), Postdoctoral fellowship Ministerio de Educacion y Ciencia (Spain) MEC/Fulbright EX 2005-0517 (A.R.).

References

Bakin RE, Gioeli D, Sikes RA, Bissonette EA and Weber MJ. (2003). Cancer Res, **63**, 1981–9.
Bargagna-Mohan P, Baek SH, Lee H, Kim K and Mohan R. (2005). Bioorg Med Chem Lett, **15**, 2724–7.
Ben-Neriah Y. (2002). Nat Immunol, 3, 20–6.
Birmingham A, Anderson EM, Reynolds A, Ilsley-Tyree D, Leake D, Fedorov Y, Baskerville S, Maksimova E, Robinson K, Karpilow J, Marshall WS and Khvorova A. (2006). Nat Methods, **3**, 199–204.
Borawski J, Lindeman A, Buxton F, Labow M and Gaither LA. (2007). J Biomol Screen, **12**, 546–59.
Bubendorf L, Kononen J, Koivisto P, Schraml P, Moch H, Gasser TC, Willi N, Mihatsch MJ, Sauter G and Kallioniemi OP. (1999). Cancer Res, **59**, 803–6.
Chandu D and Nandi D. (2004). Res Microbiol, **155**, 710–9.
Ciechanover A and Iwai K. (2004). IUBMB Life, **56**, 193–201.
Copeland RA, Pompliano DL and Meek TD. (2006). Nat Rev Drug Discov, **5**, 730–9.
Craft N, Shostak Y, Carey M and Sawyers CL. (1999). Nat Med, **5**, 280–5.
Culig Z, Hobisch A, Cronauer MV, Radmayr C, Trapman J, Hittmair A, Bartsch G and Klocker H. (1994). Cancer Res, **54**, 5474–8.

Culig Z, Klocker H, Bartsch G, Steiner H and Hobisch A. (2003). J Urol, **170**, 1363–9.

Eder IE, Hoffmann J, Rogatsch H, Schafer G, Zopf D, Bartsch G and Klocker H. (2002). Cancer Gene Ther, **9**, 117–25.

Epstein AC, Gleadle JM, McNeill LA, Hewitson KS, O'Rourke J, Mole DR, Mukherji M, Metzen E, Wilson MI, Dhanda A, Tian YM, Masson N, Hamilton DL, Jaakkola P, Barstead R, Hodgkin J, Maxwell PH, Pugh CW, Schofield CJ and Ratcliffe PJ. (2001). Cell, **107**, 43–54.

Fire A, Xu S, Montgomery MK, Kostas SA, Driver SE and Mello CC. (1998). Nature, **391**, 806–11.

Franco OE, Onishi T, Yamakawa K, Arima K, Yanagawa M, Sugimura Y and Kawamura J. (2003). Prostate, **56**, 319–25.

Gaddipati JP, McLeod DG, Heidenberg HB, Sesterhenn IA, Finger MJ, Moul JW and Srivastava S. (1994). Cancer Res, **54**, 2861–4.

Gaither A and Iourgenko V. (2007). RNA interference technologies and their use in cancer research. Curr Opin Oncol. **19**(1),50–4. Review.

Glickman MH, Rubin DM, Fried VA and Finley D. (1998). Mol Cell Biol, **18**, 3149–62.

Godoy-Tundidor S, Hobisch A, Pfeil K, Bartsch G and Culig Z. (2002). Clin Cancer Res, **8**, 2356–61.

Griffith EC, Su Z, Turk BE, Chen S, Chang YH, Wu Z, Biemann K and Liu JO. (1997). Chem Biol, **4**, 461–71.

Heidel JD, Hu S, Liu XF, Triche TJ and Davis ME. (2004). Nat Biotechnol, **22**, 1579–82.

Hoffmann J and Sommer A. (2005). J Steroid Biochem Mol Biol, **93**, 191–200.

Hon WC, Wilson MI, Harlos K, Claridge TD, Schofield CJ, Pugh CW, Maxwell PH, Ratcliffe PJ, Stuart DI and Jones EY. (2002). Nature, **417**, 975–8.

Huret J-L. (2008). *Database: http://atlasgeneticsoncology.org.*

Karin M and Ben-Neriah Y. (2000). Annu Rev Immunol, *18*, 621–63.

Kirschberg TA, VanDeusen CL, Rothbard JB, Yang M and Wender PA. (2003). Org Lett, **5**, 3459–62.

Kitchen DB, Decornez H, Furr JR and Bajorath J. (2004). Nat Rev Drug Discov, **3**, 935–49.

Lee H, Puppala D, Choi EY, Swanson H and Kim KB. (2007). Chembiochem, **8**, 2058–62.

Liao X, Tang S, Thrasher JB, Griebling TL and Li B. (2005). Mol Cancer Ther, **4**, 505–15.

Linja MJ, Savinainen KJ, Saramaki OR, Tammela TL, Vessella RL and Visakorpi T. (2001). Cancer Res, **61**, 3550–5.

Lyapina SA, Correll CC, Kipreos ET and Deshaies RJ. (1998). Proc Natl Acad Sci U S A, **95**, 7451–6.

Montironi R. (2001). BMJ, **322**, 378–9.

Nandi D, Tahiliani P, Kumar A and Chandu D. (2006). J Biosci, **31**, 137–55.

Nelson WG, De Marzo AM and Isaacs WB. (2003). N Engl J Med, **349**, 366–81.

Ohh M, Park CW, Ivan M, Hoffman MA, Kim TY, Huang LE, Pavletich N, Chau V and Kaelin WG. (2000). Nat Cell Biol, **2**, 423–7.

Parkin DM, Bray F, Ferlay J and Pisani P. (2005). CA Cancer J Clin, **55**, 74–108.

Pickart CM. (2001). Annu Rev Biochem, **70**, 503–33.

Pickart CM and Cohen RE. (2004). Nat Rev Mol Cell Biol, **5**, 177–87.

Rollins CT, Rivera VM, Woolfson DN, Keenan T, Hatada M, Adams SE, Andrade LJ, Yaeger D, van Schravendijk MR, Holt DA, Gilman M and Clackson T. (2000). Proc Natl Acad Sci U S A, **97**, 7096–101.

Root DE, Hacohen N, Hahn WC, Lander ES and Sabatini DM. (2006). Nat Methods, **3**, 715–9.

Sakamoto KM. (2005). Methods Enzymol, **399**, 833–47.

Sakamoto KM, Kim KB, Kumagai A, Mercurio F, Crews CM and Deshaies RJ. (2001). Proc Natl Acad Sci U S A, **98**, 8554–9.

Sakamoto KM, Kim KB, Verma R, Ransick A, Stein B, Crews CM and Deshaies RJ. (2003). Mol Cell Proteomics, **2**, 1350–8.

Santen RJ. (1992). J Clin Endocrinol Metab, **75**, 685–9.

Savarese DM, Halabi S, Hars V, Akerley WL, Taplin ME, Godley PA, Hussain A, Small EJ and Vogelzang NJ. (2001). J Clin Oncol, **19**, 2509–16.

Schneekloth JS, Jr., Fonseca FN, Koldobskiy M, Mandal A, Deshaies R, Sakamoto K and Crews CM. (2004). J Am Chem Soc, **126**, 3748–54.

Semple CA. (2003). Genome Res, **13**, 1389–94.

Shou J, Massarweh S, Osborne CK, Wakeling AE, Ali S, Weiss H and Schiff R. (2004). J Natl Cancer Inst, **96**, 926–35.

Sin N, Meng L, Wang MQ, Wen JJ, Bornmann WG and Crews CM. (1997). Proc Natl Acad Sci U S A, **94**, 6099–103.

Stumpfe D, Ahmed HE, Vogt I and Bajorath J. (2007). Chem Biol Drug Des, **70**, 182–94.

Tanimoto K, Makino Y, Pereira T and Poellinger L. (2000). Embo J, **19**, 4298–309.

Taplin ME, Bubley GJ, Ko YJ, Small EJ, Upton M, Rajeshkumar B and Balk SP. (1999). Cancer Res, **59**, 2511–5.

Ueda T, Bruchovsky N and Sadar MD. (2002). J Biol Chem, **277**, 7076–85.

Varshavsky A. (2005). Trends Biochem Sci, **30**, 283–6.

Voges D, Zwickl P and Baumeister W. (1999). Annu Rev Biochem, **68**, 1015–68.

Weissman AM. (2001). Nat Rev Mol Cell Biol, **2**, 169–78.

Wender PA, Mitchell DJ, Pattabiraman K, Pelkey ET, Steinman L and Rothbard JB. (2000). Proc Natl Acad Sci U S A, **97**, 13003–8.

Yang W, Rozamus LW, Narula S, Rollins CT, Yuan R, Andrade LJ, Ram MK, Phillips TB, van Schravendijk MR, Dalgarno D, Clackson T and Holt DA. (2000). J Med Chem, **43**, 1135–42.

Yeh JR, Mohan R and Crews CM. (2000). Proc Natl Acad Sci U S A, **97**, 12782–7.

Zegarra-Moro OL, Schmidt LJ, Huang H and Tindall DJ. (2002). Cancer Res, **62**, 1008–13.

Zhu H, Zhu Y, Hu J, Hu W, Liao Y, Zhang J, Wang D, Huang X, Fang B and He C. (2007). Int J Cancer, **121**, 1366–72.

Index